Introduction to the
Internet for Engineers

McGraw-Hill's BEST—*Basic Engineering Series and Tools*

Bertoline, *Introduction to Graphics Communications for Engineers*

Burghardt, *Introduction to Engineering Design and Problem Solving*

Chapman, *Introduction to Fortran 90/95*

Donaldson, *The Engineering Student Survival Guide*

Eide et al., *Introduction to Engineering Design*

Eide et al., *Introduction to Engineering Problem Solving*

Eisenberg, *A Beginner's Guide to Technical Communication*

Gottfried, *Spreadsheet Tools for Engineers: Excel '97 Version*

Greenlaw and Hepp, *Introduction to the Internet for Engineers*

Mathsoft's Student Edition of Mathcad Version 7.0

Palm, *Introduction to MATLAB for Engineers*

Pritchard, *Mathcad: A Tool for Engineering Problem Solving*

Smith, *Project Management and Teamwork*

Tan and D'Orazio, *C Programming for Engineering and Computer Science*

Introduction to the
Internet for Engineers

R<small>AYMOND</small> G<small>REENLAW</small>
Armstrong Atlantic State University

E<small>LLEN</small> H<small>EPP</small>
University of New Hampshire

 **WCB
McGraw-Hill**

Boston   Burr Ridge, IL   Dubuque, IA   Madison, WI   New York   San Francisco
St. Louis   Bangkok   Bogotá   Caracas   Lisbon   London   Madrid   Mexico City
Milan   New Delhi   Seoul   Singapore   Sydney   Taipei   Toronto

WCB/McGraw-Hill

A Division of The **McGraw·Hill** *Companies*

INTRODUCTION TO THE INTERNET FOR ENGINEERS

This book is printed on acid-free paper.

1 2 3 4 5 6 7 8 9 0 DOC/DOC 9 4 3 2 1 0 9

ISBN 0-07-229143-5

Vice president/Editor-in-chief: *Kevin T. Kane*
Publisher: *Thomas Casson*
Executive editor: *Eric M. Munson*
Marketing manager: *John T. Wannemacher*
Project manager: *Amy Hill*
Senior production supervisor: *Heather D. Burbridge*
Freelance design coordinator: *Mary Christianson*
Cover image: *The Image Bank*
Supplement coordinator: *Matt Perry*
Compositor: *Techsetters, Inc.*
Typeface: *10/12 Janson*
Printer: *R. R. Donnelley & Sons Company*

Library of Congress Cataloging-in-Publication Data

Greenlaw, Raymond
    Introduction to the Internet for engineers/Raymond Greenlaw,
  Ellen Hepp.
        p.   cm.
    Includes index.
    ISBN 0-07-229143-5
    1. Internet (Computer network).   2. World Wide Web (Information
  retrieval system).   I. Hepp, Ellen.   II. Title.
  TK5105.875.I57G744 1999
  004.67'8—dc21   98-32297

http://www.mhhe.com

To Laurel and Celeste.

To Mark, Andrew, Elisa, and Crissy.

Contents in Brief

	Preface	xv
1	Electronic Mail	1
2	Jump Start: Browsing and Publishing	37
3	The Internet	87
4	The World Wide Web	127
5	Search Topics	161
6	Telnet and FTP	187
7	HTML	201
A	Internet Service Providers	251
B	Text Editing	257
C	Pine Mail Program	263
D	Basic UNIX	273
E	HTML Tags	283
F	My URLs	289
	Glossary	299
	Bibliography	313
	Index	319

Contents

Preface   xv

1 Electronic Mail   1

1.1 Introduction   1
1.2 E-Mail: Advantages and Disadvantages   2
    1.2.1 Advantages   3
    1.2.2 Disadvantages   3
1.3 Userids, Passwords, and E-Mail Addresses   5
    1.3.1 Userids   5
    1.3.2 Passwords   6
    1.3.3 E-Mail Addresses   7
    1.3.4 Domain Names   8
    1.3.5 E-Mail Address Determinations   10
    1.3.6 Local and Systemwide Aliases   10
1.4 Message Components   13
1.5 Message Composition   16
    1.5.1 Structure   17
    1.5.2 Netiquette   19
    1.5.3 Composition   20
1.6 Mailer Features   22
    1.6.1 Compose, File, and Reply   23
    1.6.2 Bracketed Text and Include   23
    1.6.3 Forwarding   25
1.7 E-Mail Inner Workings   26
    1.7.1 Mailer, Mail Server, and Mailbox   26
    1.7.2 Store and Forward Features   27
    1.7.3 Central Mail Spool and IMAP   29
    1.7.4 Bounce Feature   30
1.8 E-Mail Management   32
    1.8.1 Action Options   32
    1.8.2 Vacation Programs   33
    1.8.3 E-Mail and Businesses   34
1.9 MIME Types   35

2   Jump Start: Browsing and Publishing   **37**

2.1   Introduction   37
2.2   Browser Bare Bones   37
      2.2.1   Browser Window Terminology   38
      2.2.2   Menu Bar   39
      2.2.3   Toolbar   40
      2.2.4   Hot Buttons   42
      2.2.5   Hyperlinks   43
2.3   Coast-to-Coast Surfing   43
      2.3.1   Web Terminology   44
      2.3.2   Uniform Resource Locator (URL)   45
2.4   HyperText Markup Language: Introduction   47
      2.4.1   HTML Tag Syntax   48
      2.4.2   HTML Document Creation   49
2.5   Web Page Installation   52
      2.5.1   Basic Principles   53
      2.5.2   A Specific Example   53
2.6   Web Page Setup   56
      2.6.1   Head Tag   56
      2.6.2   HTML and Colors   62
      2.6.3   Body Tag   64
      2.6.4   HTML Font Colors   68
      2.6.5   Font Size   68
      2.6.6   Font Face   69
      2.6.7   HTML Comments   70
2.7   HTML Formatting and Hyperlink Creation   73
      2.7.1   Paragraph Tag   74
      2.7.2   Heading Tags   75
      2.7.3   Anchor Tag   77
      2.7.4   Image Tag   82

3   The Internet   **87**

3.1   Introduction   87
3.2   The Internet Defined   88
      3.2.1   The Information Superhighway   88
      3.2.2   Interesting Internet Facts   89
3.3   Internet History   90
      3.3.1   1960s Telecommunications   91
      3.3.2   1970s Telecommunications   91
      3.3.3   1980s Telecommunications   92
      3.3.4   1990s Telecommunications   93
      3.3.5   Internet Growth   95

3.4 The Way the Internet Works   96
   3.4.1 Network Benefits   96
   3.4.2 Interconnected Networks and Communication   97
   3.4.3 Physical Components   99
   3.4.4 Network Connections   99
   3.4.5 Client-Server Model   100
   3.4.6 IP Addresses   100
   3.4.7 Internet Protocol Version 6 (IPv6)   103
   3.4.8 Web Page Retrieval   103
3.5 Internet Congestion   105
   3.5.1 World Wide Wait Problem   105
   3.5.2 Technical Solutions   106
   3.5.3 Issues and Predictions   107
3.6 Internet Culture   108
   3.6.1 Critical Evaluation of Information   109
   3.6.2 Freedom of Expression   109
   3.6.3 Communication Mechanisms   111
   3.6.4 Advertising   112
   3.6.5 Societal Impact   113
3.7 Business Culture and the Internet   114
   3.7.1 On-Line Businesses   115
   3.7.2 Three Sample Companies   116
   3.7.3 On-Line Business Hurdles   118
   3.7.4 Cookies   118
   3.7.5 Business and Safety/Security on the Web   119
   3.7.6 Legal Environment   119
   3.7.7 U.S. Government's Commitment to Electronic Commerce   120
3.8 Collaborative Computing and the Internet   121
   3.8.1 Collaborative Computing Defined   121
   3.8.2 Applications   122
   3.8.3 Impact   123
   3.8.4 Future Prospects   124

4 The World Wide Web   127
4.1 Introduction   127
4.2 The Web Defined   127
4.3 Miscellaneous Web Browser Details   129
   4.3.1 Personal Preferences   129
   4.3.2 Bookmarks   130
   4.3.3 Plug-ins and Helper Applications   133
   4.3.4 Web Browsers Comparison: Netscape and Microsoft   134

4.4 Web Writing Styles   136
    4.4.1 The Biography   137
    4.4.2 The Business Exposition   139
    4.4.3 The Guide   141
    4.4.4 The Tutorial   141
    4.4.5 Writing Genres Summary   144
4.5 Web Presentation Outline, Design, and Management   145
    4.5.1 Goal Setting   146
    4.5.2 Outlining   148
    4.5.3 Navigating   153
    4.5.4 Designing and Coding   156
    4.5.5 Revising   157
4.6 Registering Web Pages   159

5 Search Topics   161
5.1 Introduction   161
5.2 Directories, Search Engines, and Metasearch Engines   162
    5.2.1 Directories   162
    5.2.2 Popular Directories   163
    5.2.3 Search Engines   163
    5.2.4 Popular Search Engines   165
    5.2.5 Metasearch Engines   165
    5.2.6 Popular Metasearch Engines   165
        Ellen and Ray's Choices   166
    5.2.7 White Pages   166
    5.2.8 Popular White Pages   166
5.3 Search Fundamentals   167
    5.3.1 Search Terminology   169
    5.3.2 Pattern Matching Queries   170
    5.3.3 Boolean Queries   172
    5.3.4 Search Domain   173
    5.3.5 Search Subjects   174
5.4 Search Strategies   175
    5.4.1 Too Few Hits: Search Generalization   175
    5.4.2 Too Many Hits: Search Specialization   176
    5.4.3 Sample Searches   176
5.5 How Does a Search Engine Work?   178
    5.5.1 Search Engine Components   178
    5.5.2 User Interface   179
    5.5.3 Searcher   179

	5.5.4	Evaluator	179
	5.5.5	Gatherer	180
	5.5.6	Indexer	184
	5.5.7	Summary	184

6 Telnet and FTP   **187**

6.1 Introduction   187
6.2 Telnet and Remote Login   187
   6.2.1 Telnet   188
   6.2.2 Remote Login   191
6.3 File Transfer   192
   6.3.1 Graphical File Transfer Clients   193
   6.3.2 Text-Based File Transfer Clients   195
   6.3.3 File Compression   196
   6.3.4 Anonymous File Transfer   196
   6.3.5 Archie   197
6.4 Computer Viruses   198
   6.4.1 Definitions   198
   6.4.2 Virus Avoidance and Precautions   199

7 HTML   **201**

7.1 Introduction   201
7.2 Semantic Versus Syntactic Based Style Types   201
   7.2.1 Semantic Based Style Types   202
   7.2.2 Syntactic Based Style Types   207
   7.2.3 Style Type Usage   210
7.3 Headers and Footers   211
   7.3.1 Headers   211
   7.3.2 Horizontal Rule Tag   213
   7.3.3 Footers   214
7.4 Lists   217
   7.4.1 Ordered Lists   217
   7.4.2 Unordered Lists   219
   7.4.3 Definition Lists   221
   7.4.4 Nested Lists   223
7.5 Tables   229
   7.5.1 Table Usage   229
   7.5.2 HTML Table Tags   230
   7.5.3 Frequently Asked Questions   244
7.6 Debugging   247
7.7 More Advanced Topics   250

A Internet Service Providers   251

B Text Editing   257

C Pine Mail Program   263

D Basic UNIX   273

E HTML Tags   283

F My URLs   289

Glossary   299
Bibliography   313
Index   319

Preface

The Internet has experienced spectacular growth over the last few years. A wide range of knowledge is needed by anyone interested in publishing on and participating in the *World Wide Web*. Many would argue that all members of society should have a basic understanding of computing principles and the capability to track down information on the Web. In other words, everyone should be *Internet literate*. The amount of information available on-line is so vast that anyone interested in obtaining timely news, stock updates, a hard-to-find product, or basic research information cannot overlook the Web. All scientists and engineers need to be familiar with the information contained in this book.

The level of sophistication for which we are aiming in this book is not the "point and click" level, nor the "hacker" end of the spectrum. We are interested in helping you learn enough that you are comfortable performing the following functions (among others):

- Sending and receiving electronic mail (e-mail).

- Browsing the World Wide Web.

- Publishing on the Web.

- Coding in HyperText Markup Language (HTML).

- Using search engines.

- Processing on-line information in a critical fashion.

- Using such Internet applications as Telnet and FTP.

- Developing a good grasp of computer terminology and acronyms.

Organization of the Text

The material in this text is organized for a short course. It is designed to give the reader a quick start to using the Internet and the World Wide Web. For a more comprehensive treatment of this and additional material, the reader should consult *In-line/On-line: Fundamentals of the Internet and the World Wide*

Web, published by McGraw-Hill. This book contains numerous exercises at the end of each section. It can also be used as a self-study guide for anyone interested in recent computing developments revolving around the Internet.

The book chapters deal with the following topics: e-mail, Web browsing, Web page installation, the Internet, the World Wide Web, search tools, Telnet and FTP, and HTML. The appendices provided deal with more specialized issues, such as Internet Service Providers, text editors, mailers, and operating systems. In addition, we provide a list of references, a list of all HTML tags presented in the book, a glossary, a place for you to record your own URLs, and an index.

Icons are used in the margin of the text to highlight important or interesting points and to suggest activities to the reader.

Denotes an interesting tidbit of information, a factoid.

Notes a hint.

Marks extensive or important information.

Go on-line and experiment using the commands described in the text.

Flags a recent news item.

Surf the Web to experiment with the topic explained in the text.

Accompanying Web Presentations

Two Web presentations are associated with the book *In-line/On-line: Fundamentals of the Internet and the World Wide Web* and may be used with this book as well: one is called "class" and the other is called "book." The class presentation can be customized by an instructor and includes the following material: assignments, frequently asked questions, grading information, a hall of fame, information for parents, a project outline, a student directory, a syllabus, and a "welcome" message.

The book presentation contains the following items: a set of lecture summaries, a collection of useful links for each chapter, additional examples (not contained in the book) that utilize many HTML tags, HTML code for all of the "screen shots" contained in the book, search engine links, sample quizzes, updates of recent material/new developments, and corrections.

The on-line presentations can be accessed through McGraw-Hill's Web presentation:

www.mhhe.com

or by visiting one of our Web pages.

Legal and Ethical Guidelines

You will want to read on-line information, review paper handouts, and discuss local policies with your system administrator. The standard rules follow common sense. Let us express the guidelines in the form of a "do not" list. Do not:

- Copy, borrow, or steal another person's work.

- Try to break into another person's account.

- Forge e-mail.

- Steal passwords.

- Make selfish use of system resources.

- Produce offensive material.

- Violate computer policies.

In summary, exercise good judgment and remember that using computer facilities is considered a privilege, not a right.

Suggestions and Corrections

The text may contain some errors, and certain topics that readers feel are especially relevant may have been omitted. In anticipation of possible future printings, we would like to correct our mistakes and incorporate as many suggestions as possible. Please send comments to us via e-mail at:

emhepp@aol.com   or   greenlaw@pirates.armstrong.edu

Acknowledgments

Thanks to Jim Cerny and others who have commented on drafts of this material.

Thanks to the students at the University of New Hampshire who enrolled in Computer Science 403 during the academic years 1996–97 and 1997–98. Their comments and suggestions have helped to make this a better book. Thanks to Betsy Jones, Emily Gray, and Amy Hill for their excellent and supportive work on this material.

Ray Greenlaw
Ellen Hepp

Introduction to the
Internet for Engineers

Electronic Mail

1

1.1 Introduction

Electronic mail, or *e-mail* as it is most commonly referred to, is the subject of this chapter. Our goals are to acquaint new users with the fundamental principles of e-mail, to reinforce these basics for intermediate users who have been working with e-mail for a few years, and to point out a few subtleties to more experienced users. The following topics, among others, are discussed in this chapter:

OBJECTIVES

- Basic e-mail facts.

- E-mail advantages and disadvantages.

- E-mail addresses, passwords, and userids.

- Message components.

- Message composition.

- Mailer features.

- E-mail inner workings.

- E-mail management.

- Multipurpose Internet Mail Extensions (*MIME*) types

   E-mail infiltrates many areas of the Internet, which is one of the primary motivations for including this topic at the beginning of this text. You will find uses of e-mail throughout the book. Since there are so many different *e-mail programs* (synonyms include *mailers*, *mail clients*, and *mail applications*), we will stick to a generic description of e-mail. (The Pine mail program is covered in detail in Appendix C.) The basic concepts carry over directly to most mailers. It is important to grasp the fundamentals of e-mail, since new and improved e-mail programs are continuously being introduced. Additionally, if you switch companies at some point in your career, there is a good chance your new employer will use a different e-mail system than the one you are currently using.

   Mail clients are programs that are used to manage, read, and compose e-mail. These programs have become very sophisticated in the last few years. This has made using e-mail much easier and more efficient. A few popular mail clients are Elm, Eudora, mail-tool, mailx, Microsoft Exchange, and Pine. Most *Web browsers* come with a built-in mail application, although at present these are not as sophisticated as some stand-alone mail clients. Many *Internet Service Providers* (ISPs) provide their own mail clients. (Appendix A discusses how to select an ISP.)

   E-mail is one of the most popular services available through the Internet. In the early days of the Internet,[1] e-mail emerged as an inexpensive and efficient means of communication between researchers, scientists, people in high-tech jobs, and those in academia. The words "inexpensive" and "efficient" are used here in relation to a telephone call; e-mail is a lot less expensive than a phone call and nearly as fast. And, like postal mail (*snail mail* or *s-mail* as it is sometimes called now), a message can easily be printed to provide concrete documentation of the correspondence. Furthermore, from the user point of view, sending a 100-page document requires no more effort or cost than sending a one-page document: both documents will typically have the same delivery time. There are some mail systems that cannot handle very large messages, in which case the user might have to find an alternative way of getting the information to the intended recipient.

1.2 E-Mail: Advantages and Disadvantages

Today many people all over the world have been exposed to e-mail—they have either heard of it, used it occasionally, or felt they could not function without it. E-mail began as a system in which an individual user could send a plain-text message via the Internet to another user. E-mail has grown in ways that no one predicted. In contrast to the simplicity of e-mail functions a short while ago, people can now receive and send e-mail to:

---

[1] Prior to 1990.

- Nearly any country in the world.

- One of millions of computer users.

- Many users at once.

- Computer programs.

This last point is worth elaborating. Examples in which sending a message directly to a computer program proves useful include subscribing to a mailing list, submitting a paper or form electronically, and accessing remote Internet services such as *file transfer* and *gopher*.

1.2.1 Advantages

As in the early days of the telephone, the original users of e-mail only had a limited number of people with whom they could communicate. Now that e-mail is more prevalent, some of the advantages of using it are clear.

Convenience—There are no trips to the post office, and no need to search for stationery and stamps. Sending a memo or short note is easy. A message can be informal or formal. E-mail makes publishing and discussing very easy, for example, in the forms of mailing lists and newsgroups.

Speed—E-mail is fast, based on the speed of the underlying communication network.

Inexpensive—Once you are on-line, the cost of sending a message is small.

Printable—A hard copy is easy to obtain. However, since a great deal of correspondence does not need to be printed, using e-mail saves on natural resources. You can keep an electronic copy of a message for your own records.

Reliable—Although messages are occasionally lost, this is rare. Many mail systems will notify the sender if an e-mail message was undeliverable.

Global—Ever increasingly, people and businesses all over the world are using e-mail.

Generality—E-mail is not limited to text; it allows the transfer of graphics, programs, and even sounds.

1.2.2 Disadvantages

Despite all of the advantages, we should bear in mind that not everyone everywhere has access to e-mail. Although the telephone is not truly universal either, it still far outdistances e-mail in terms of its worldwide availability.

Misdirection—With e-mail, you are your own worst enemy. It is far more likely that you will accidentally send e-mail to an unintended recipient than it is for someone actually to intercept your e-mail.

Interception—It is possible, although unlikely, that eavesdroppers are "listening in" on e-mail correspondence. As a rule of thumb, never send an e-mail message that you would not want the whole world to see. It is simple for someone to pass on your message, called e-mail *forwarding*, to another party.

Forgery—E-mail does not preclude forgeries, that is, someone impersonating the sender, since the sender is usually not authenticated in any way.

Overload—E-mail can also be too convenient and result in a flood of mail.

Junk—Another more recent negative development involves "junk e-mail," or unsolicited commercial e-mail. This flooding of undesirable or inappropriate e-mail is sometimes referred to as *spam*[2] and is becoming a serious problem. Some on-line services provide help in dealing with this unwanted e-mail, and there are Web sites that suggest strategies for coping with spam. The process of sending junk e-mail to lots of sites simultaneously is known as *spamming*. Please avoid this type of abuse of the e-mail system.

No response—A mild frustration sometimes associated with using e-mail is dealing with recipients who do not read and respond to their e-mail on a regular basis. However, this occurs using regular postal mail as well. There are programs that can be used on some systems to check when a person last received an e-mail message. Although e-mail is highly reliable, do not immediately assume that a lack of a response from someone indicates a negative reaction. They[3] may not have received your message, or they may not have read their mail.

Despite the disadvantages, which are for the most part the same as those associated with any communication mechanism, many people prefer using e-mail to either using the telephone or writing a letter. E-mail is less formal than a letter and not as intrusive as a phone call. An e-mail response can be pondered and formulated, if necessary, or sent back right away. On the whole, the advantages of e-mail are great, and the disadvantages, although real, are acceptable when compared with alternatives.

---

[2] As an historical note, we point out that Spam was a meat product originally fed to soldiers in World War II. Spam is commercially available and celebrated its 50th birthday in 1997. Spam has a cult-like following. The use of *spam* for e-mail was taken from the famous Monty Python skit in which John Cleese (customer) asks Eric Idle (waitress) what they have on the menu for breakfast. "She" then rattles off endless combinations of "eggs and Spam," "Spam, eggs, and Spam," and so on to the point that everyone is finally ready to scream if they hear the word again.

[3] In this book, we use the pronouns "their," "them," and "they," even where the correct usage is singular. This seems to be a better solution than: (1) using he/she, (2) alternating the gender of pronouns throughout the book, or (3) using "she" to compensate for years of overuse of masculine pronouns.

EXERCISES 1.2 E-Mail: Advantages and Disadvantages

1. Write a paragraph or two aimed at someone who was around before the advent of computers. Describe your idea about e-mail for the future, including how and why you see it becoming popular and universal.

2. Write a paragraph explaining your vision of the future of e-mail.

3. Describe one good and one bad experience that you or someone you know has had with e-mail.

4. What is an *e-mail bomb*?

5. Have you ever received spam? If so, what was the nature of the message?

1.3   Userids, Passwords, and E-Mail Addresses

Before continuing with our e-mail discussion, it is worth devoting one section to *userids*, *password* selection, and *e-mail addresses*. This section should be particularly useful for new computer users.

The paramount issue for you is to get access to a computer system. You may already have an Internet Service Provider and use e-mail, or perhaps you are a student who has an account on a campus computer. The important items you must know to access a computer system are your *userid* or *login name* and your *password*.

1.3.1   Userids

Synonyms for userid are *user name* and *account name*. Userid is merely the concatenation of the word "user" and the abbreviation "id," standing for identification. Your userid identifies you to the computer.

In most settings, userids have some mnemonic meaning. For example, Ponette Beth Lucas' userid might be her first two initials joined by her last name, pblucas, while Nick Michael Walters' userid might be his last name, walters. Such userids are much easier to remember than something like P13245. Very few people will be able to put a name together with P13245. If you have a choice, pick as descriptive a name as possible, but one that is also easy to type and associate with you. In some cases, these goals are mutually exclusive, but keep them in mind.

If your name were Mary H. Lamb, sensible login names would be mhlamb, MaryLamb, mlamb, or lamb. Note that uppercase or lowercase is normally not significant in e-mail names. That is, MaryLamb and marylamb are treated the same. However, one is easier to read and maybe shows a bit more respect for the account owner. If your name is Steve Village, do not pick a user name of Steve, since there are several million Steves in the world, and such a userid by itself would not uniquely identify you.

Internet Service Providers sometimes perform the unfortunate practice of handing users their next available letter or number combination as a userid. Their behavior is akin to that of a Division of Motor Vehicles. When you obtain a license plate for your car, the sequence of symbols you get for your plate depends only upon when you arrive at the counter and sign your check. What we recommend is that, if possible, you get a "vanity plate" for your userid. Also, many users keep the same userid for years, so it is worthwhile selecting a good one the first time around.

If you have a common name or are part of a large organization, your first name and last name combined will probably not uniquely identify you. In such cases, it may be necessary to append a number to your name. For example, `SusanSmith14` identifies the fifteenth Susan Smith on a system (assuming `SusanSmith` or `SusanSmith0` identifies the first).

1.3.2 Passwords

Your *password* is a secret code that *authenticates* you to the computer. This is done simply to check that you are who you say you are. In theory, you are the only one who knows the password to your computer account, and no one except you should be able to log in to your account.

On most computer systems, your password will have to meet basic criteria in order to be allowed. That is, the computer system requires these conditions to be met as a security measure. After all, if your password is a word in the dictionary, a programmer might be able to write a program that tries to break into your account by simply testing every word in the dictionary as a password. Since dictionaries are easily accessible on-line, this is not as hard as it might sound.

A good password should:

- Be at least five characters long.

- Contain a nonalphabetical symbol such as &, %, or !.

- Contain a number.

- Possess uppercase and lowercase letters.

Case is significant in passwords. Hard-to-guess passwords are `OG&$jaNp`, `MuL()a#`, and `v*p*!CA`. The drawback to these examples is that they are also hard to remember. One school of thought is if you are going to be logging into a multiuser system, pick fairly easy-to-remember passwords and change them regularly. Another alternative is to set a good password and stick with it.

You should change your password immediately if you think it has been compromised. On most systems, changing a password involves running a program, typing in the old password, typing in the new password, and finally typing in the new password a second time for verification. When you type a password, it is not echoed on screen. Modern systems usually keep an individual history file on passwords, so that you cannot quickly reuse or switch back to a previous password if the system forces you to change passwords. Avoid writing down passwords, especially in on-line files.

1.3.3 E-Mail Addresses

The basic form of an e-mail address is

```
username@hostname.subdomain.domain
```

There are some exceptions, but this format covers most common addresses. The text before the @ (pronounced "at") sign[4] specifies the `username` of the individual, while the text after the @ sign indicates how the computer system can locate that individual's *mailbox*. We have already covered userids, so let us focus on the suffix following @.

Consider a few sample e-mail addresses before we make any general statements. Take for example:

```
maria@cs.colorado.edu
```

The first important point to be aware of is that the suffix following the @ sign gets more general from left to right. That is, `cs` is a subdomain of `colorado`, `colorado` is a subdomain of `edu`, and `edu` specifies a (*generic*) *top-level domain name*. In this case, Maria is a computer science major at the University of Colorado. If `maria` had her own computer called `tennis`, the e-mail address

```
maria@tennis.cs.colorado.edu
```

would also work.

The number of periods (a period is pronounced "dot") varies from e-mail address to e-mail address. Most addresses have either one or two dots. For example, consider the following e-mail address:

```
mark@wheelabrator.com
```

This address has a top-level domain of `com`. The `com` stands for "commercial." The subdomain `wheelabrator` is a commercial entity, and `mark` is a computer user who works at Wheelabrator.

---

[4] The @ was selected in July 1972 by Ray Tomlinson at Bolt, Beranek, and Newman, Inc. According to *Where Wizards Stay Up Late*, Hafner and Lyon, page 192: "I got there first, so I got to choose any punctuation I wanted," Tomlinson said. "I chose the @ sign." The character also had the advantage of meaning "at" the designated institution. He had no idea he was creating an icon for the wired world.

A given field in an e-mail address, that is, a part separated by dots, can be no more than 63 characters long. All fields combined must total less than 256 characters.

How are e-mail addresses read out loud? This is important to know so that you can communicate your e-mail address and can record someone else's. The first address example is read as "maria at c s dot colorado dot e d u," and the second one is read as "mark at wheelabrator dot com."

1.3.4 Domain Names

Currently only a small number of top-level domain names exist in the United States, and they are listed in Table 1.1. The big seven generic top-level domain names in the United States are com, edu, gov, int, mil, net, and org. In addition, every country has its own top-level domain name. In the United States, these are called *country codes*, a sampling of which are shown in Table 1.2. It is easy to track down a listing of all country codes on-line.

There are proposals to expand the number of top-level domain names. The decision involves economics and politics as much as technology. The number of top-level domain names is an artifact of the early years of the Internet. In total, including all country codes, there are about 150 top-level domain names.

Figure 1.1 provides a convenient way of viewing the organization of the domain names. In the figure, we show a very small fragment of the *domain name space* (DNS). It is important to realize that this is a *distributed naming scheme* that follows the boundaries of countries and organizations, rather than those of networks. The DNS represents a key feature of the success of the Internet: distributed assignment and recordkeeping of a database, so that a local authority (and its backup) is responsible for local information. This means that administrators do not face the impossible task of trying to maintain one enormous central database that would be a choke point and would strangle Internet use.

TABLE 1.1 The Generic Top-Level Domain Names Used in the United States.

Domain Name	Meaning
com	commercial business
edu	education
gov	U. S. government agency
int	international entity
mil	U. S. military
net	networking organization
org	nonprofit organization

TABLE 1.2 A Small Sampling of Country Top-Level Domain Names.

Domain Name	Meaning
au	Australia
ca	Canada
cl	Chile
de	Germany
es	Spain (España)
fr	France
hk	Hong Kong
jp	Japan
nl	The Netherlands
uk	United Kingdom
us	United States
za	South Africa

FIGURE 1.1 A Small Fragment of Domain Name Space.

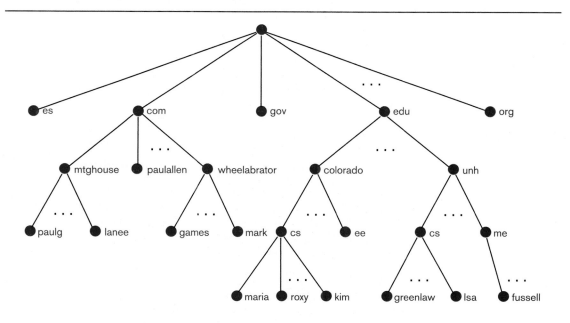

Imagine Figure 1.1 as an inverted *tree*, with the top circle being the root. Traversing a path from the root to a leaf corresponds to going from right to left in an e-mail address. For example, the user kim would have an e-mail address of

```
kim@cs.colorado.edu
```

1.3.5 E-Mail Address Determinations

Once you know the basic principles behind e-mail addresses and you have the picture of Figure 1.1 in mind, it should be fairly easy for you to guess someone's e-mail address. For example, suppose you know that Frank Denellio has his own company called Hoopsters. An educated guess as to his e-mail address would be

```
denellio@hoopsters.com
```

Since you know his company is not a nonprofit, you guess com instead of org. Similar principles and a speckle of logic can be used to make informed guesses about anyone's e-mail address. For those times when you cannot locate someone, a number of different methods are available for determining someone's e-mail address.

- Ask the person directly.

- Use a program specially designed for locating people. This usually works best for finding network or site support staff.

- Go through your browser. For example, Netscape Navigator includes a "People" button that provides access to a group of *white page directories*. White pages are like on-line phone books.

- Use a program such as *finger* to verify a guessed address. But be aware that as a security measure, many sites disallow external fingering.

- Use a *search engine*, such as *Yahoo!*, and submit a query on that person's name.

- Use a search engine to determine the primary *Web server* for the site or organization where the person is located. Then look for directory information on the site's Web pages by using a search option at that site.

1.3.6 Local and Systemwide Aliases

Having determined an individual's e-mail address, you may find that the e-mail address is too cumbersome to remember. In that case, an *alias* for the address can be set up within your own e-mail system. An alias is an easy-to-remember name that you create. The mail client associates the alias with a particular e-mail address. Each mailer has its own procedure for setting up aliases, but they are similar and are typically called something like "address book" or "nickname feature" by the mailer. For example, if your mother has an e-mail address

`LoisLane@aol.com`

and you frequently send her e-mail, you may want to set up an alias, such as `MOM`. Then, every time you want to send e-mail to your mother, you would select the address `MOM` and the mailer would insert your mother's e-mail address (or make it available for you to insert). If you have to type in the address, typing `MOM` clearly is much faster. Conceptually,

`MOM` gets replaced by `LoisLane@aol.com`

An e-mail alias can also be established for a whole group of people to whom you need to send the same message. For example, you may want to send a message to members of a class or a special interest group. To send a message to everyone in your bowling club, you might select an alias of `club` and list all your club members' e-mail addresses. Then simply addressing your e-mail to `club` will result in everyone receiving a copy of the message. Obviously, this is very convenient, because you do not have to look up the e-mail addresses of everyone who is in the club. This type of alias is sometimes referred to as a *distribution list* or *private distribution list*. The aliases we have been discussing only affect the e-mail that you send; such aliases are known as *local* or *private aliases*. Someone else's alias of `club` might include the members of their softball team.

In addition to setting up a local alias, a systemwide or *public alias* can also be established. These are aliases that are usable by everyone on any system that can send e-mail. The primary use of a public alias is to buffer the owner and their correspondents against change. The public alias should be selected to be as stable as possible and to point to the real delivery address, which can be expected to change over time. A secondary use of a public alias is to make it easy or convenient to locate someone. For example, the system administrator might decide that it would be helpful to use an alias, such as `bill`, for himself, so that anyone on the system can easily mail him questions and comments. Even though you personally do not have `bill` designated as an alias, the mailer will determine that a message sent to `bill` is a public alias and will direct your message to the correct address.

One question is, What happens if you have already designated a particular alias as one of your local aliases? Assume that you have established `bill` as a local alias for your college roommate Bill Patterson. The client will check your list of nicknames and make an appropriate substitution if one exists. Otherwise, the client will take the name at face value and send the message on its way, leaving it to other parts of the mail system to discover if the address is valid and where to deliver the message. Since, in this case, `bill` is defined locally, your e-mail to `bill` will wind up being sent to `bill@biology.ucla.edu` and not to the system administrator.

Sometimes, it is useful for the system administrator to set up public aliases for you. This way, if someone is taking an educated guess at your e-mail address, chances are that it will match one of your system aliases. For example, Bill Patterson might be assigned e-mail aliases of `billp`, `billpatterson`, `bp`, and `patterson`.

When a system administrator establishes a global alias for a list of addresses, the result is a private distribution list that allows a user to send a message to a whole group of people. To send messages to each person on the list, the sender addresses their e-mail to the name of the list. Instead of the message being received by an individual, who would then forward the message, the mailing software automatically distributes the message to each name on the list.

EXERCISES 1.3 Userids, Passwords, and E-Mail Addresses

6. What user names would you suggest for the following rock stars:

    Christina Amphlett, Mick Jagger, Jewel, Natalie Merchant, Alanis Morissette, biggie small, Steven Tyler, and Eddie Vedder?

7. (This problem requires discrete mathematics.) If you are allowed to have passwords consisting of only five lowercase letters, how many possible passwords are there? What if we also allow uppercase letters in combination with lowercase letters? Now, suppose we allow digits (0–9), as well as uppercase and lowercase letters. How many different passwords are possible of lengths five, six, and seven characters? (Factoid: Some operating systems allow passwords of up to 31 characters long.)

8. Would "concertpromoter" be a good password for a concert promoter? Explain.

9. A computer *algorithm* (which is really just a set of specific rules) for automatic userid generation follows these sequential rules:

    (a) Last name, or

    (b) First initial, last name, or

    (c) First, middle (if there is one), and last initial, or

    (d) First, middle (if there is one), and last initial, followed by a counter that starts at 2.

    Note that all account names are lowercase. If the following users were the first people added to the system in sequence, what userids would they have: John Allen, Eleanor Allen, Kendra Allen, Shirley Allen, John Allen, John K.

Allen, Johnny Allen, Joseph Allen, Joseph Mike Allen, Mark Allen, Marcus Allen, Robert John Allen, Jill Kendra Allen? Do you see any problems with the automatic userid generating algorithm? Why do you suppose computer operators use such algorithms?

10. On your computer system, what is the *default password* a new user is assigned? How long do you have before it must be changed? How frequently must the password be changed? Are there any rules you must follow in choosing a password?

11. This exercise should make you aware of the number of countries currently using e-mail. Put together a list of *all* top-level country code domain names, not just those listed in the text. Are they all two letters long? How many different countries could you locate?

12. List three additional methods (not given in the text) of tracking down a user's e-mail address.

13. A mechanical engineering professor named Barry Mark Messer works at the University of Arizona. What are two informed guesses as to what his e-mail address might be?

14. Test out any programs available to you for tracking down an e-mail address. Use them to locate an old acquaintance. Report your findings.

1.4 Message Components

You may have already been using e-mail for awhile and have a pretty good handle on the basics. Nevertheless, it is worth considering what comprises an e-mail message. Hopefully, a few items we touch on will be new to you. Let us begin by looking at a sample e-mail message. Of course, in most e-mail clients, before you see the message, you see a list of the messages showing

- Date.
- Sender name.
- Size (*bytes*).
- Subject line (usually truncated).

Sometimes, additional symbols are used to flag whether or not you have already viewed the message (see Figure 1.4). Our goal here is to explain a message's different components. Figure 1.2 depicts a sample e-mail message.

FIGURE 1.2 A Sample E-Mail Message.

Header
```
From: Alex Diaz <alex@eng.ephs.edu> Wed Jun 18 11:00 EDT 1997
Date: Wed, 18 Jun 1997 11:00:46 -0400 (EDT)
To: shiller@aol.com
Subject: bean dip
Cc: wong@sport.middlebury.edu
```

Greeting
```
Hi Guys,
```

Text
```
   Someone accidentally finished off the black bean dip last
night.  Can one of you pick up another case of it on your
way home?  I think Luke is on his bike today so you might
have to, Tak.
```

```
                        --Alex


**************************************************************
    Alex T. Diaz              |  office:    (401) 437-2134
    332 Toast Lane            |  messages:  (401) 437-0012
    East Providence,          |  fax:       (401) 437-2137
    Rhode Island 02915        |  alex@eng.ephs.edu
**************************************************************
```

Signature

Body

The first five lines of the message are referred to as the *e-mail header*. Each mail client will display slightly different header information. Often, header information is part of the e-mail message, but the mail client may not be set to display that information. Sometimes you can see these extra lines if you save the message in a mail client folder and then look at the file with an ordinary text editor or just print the message out. The header we have shown is actually an abbreviated e-mail header. The *full header* includes some additional information, such as parts of the route the message took to reach your computer and the unique *message id* associated with this particular message.

While most parts of the message are self-explanatory, we will mention them briefly to familiarize you with the terminology.

The *From* field indicates who sent the message and when. In this case Alex Diaz, whose address is alex@eng.ephs.edu, sent the message on Wednesday, June 18, 1997, at 11 am Eastern Daylight Time (EDT). Time is represented using a 24-hour clock.

The *Date* field repeats the date and includes an interesting feature: the −0400. This tells us that EDT is four hours behind Greenwich Mean Time (GMT). Greenwich, England, is the location where standard time is kept. Since e-mail is sent throughout the world, a reference by the mailer to GMT lets us deduce when the user sent the message in relation to our local time. In this example, the message was sent at 3 pm GMT. If the message had been sent from Claremont, California, instead of East Providence, Rhode Island, we would see a reference to Pacific Daylight Time (PDT) rather than EDT. In general, the *Date* field will include a reference to the time zone the sender is in.

The *To* field specifies to whom the message was sent. In this case, the recipient is `shiller@aol.com`.

The *Subject* field provides a hint as to what the message is about. Here, after only seeing the *Subject* field, we can assume the message will have something to do with "bean dip."

The *Cc* field tells us that the message was "carbon copied" to another user. Long ago when a duplicate message needed to be sent, carbon paper was used to generate the extra copy, hence the term "carbon copy" for a duplicate message. Those familiar with business letters will often see a "cc:" at the bottom, followed by the names of other recipients of the letter. In our example, the message was also delivered to the e-mail address

```
wong@sport.middlebury.edu
```

One field that does not appear that is worth mentioning is *Bcc*, which stands for *blind carbon copy*. Additional copies of the message may have been sent out. If the Bcc feature was used, we would not see it in the heading. *Bcc* is used when you do not want one or more of the recipients to know that someone else was copied on the message.

The opening

```
Hi Guys,
```

is called the *greeting* of the message. More formal messages are addressed like off-line letters and usually begin with Dear.

The main content is called the *text* of the message.

```
Someone accidentally finished off the black bean dip last
night.  Can one of you pick up another case of it on your
way home?  I think Luke is on his bike today so you might
have to, Tak.
```

The final part of the message is known as the *signature*.

```
                        --Alex


**************************************************************
    Alex T. Diaz            |  office:    (401) 437-2134
    332 Toast Lane          |  messages:  (401) 437-0012
    East Providence,        |  fax:       (401) 437-2137
    Rhode Island 02915      |  alex@eng.ephs.edu
**************************************************************
```

In many business situations involving frequent message exchanges, it is standard to omit the greeting and signature altogether.

The greeting, text, and signature form the *body* of the message. Most e-mail clients recognize the header and body divisions of e-mail messages. A third part of some messages is a *MIME attachment*. MIME is described in Section 1.9.

EXERCISES 1.4 Message Components

15. Print out the full header of an e-mail message that you composed and sent to yourself.

16. Set up an e-mail alias for yourself using your mail client. Send a test message to yourself using the alias. What name appears in the *From* field? How about the *To* field? Is the alias automatically expanded by the mailer?

17. Print out an e-mail message you received and label the different parts.

18. Does your mail client support the *Bcc* feature?

19. If you *Cc* yourself on a message twice, do you get two copies or only one?

20. Print out the most interesting and the "ugliest" signatures you have ever encountered in an e-mail message. Compare and contrast them.

21. Will your e-mail program allow "random" signatures? Explain what a random signature is.

1.5 Message Composition

The manner in which you compose an e-mail message may vary from one mail program to another. However, the basic elements remain the same, even if you are composing your e-mail outside the mailer, using a simple *text editor*.

1.5.1 Structure

If you are composing an e-mail message within a mailer, it will "prompt" you for certain information. Let us begin our discussion from the point at which you have selected the "compose" button or command. Figure 1.3 shows a typical template that a mailer might provide.

The mailer's first field is generally the *To* field. Here you should enter the e-mail address of the person to whom you are sending the message. Rather than going to the *Cc* field to enter other e-mail addresses, many mailers allow you to enter a list of names, separated by commas, on this line. This is one way to make everyone feel equally important. For example,

```
To: Joe@waterworks.com, Martha@glasses.com,
    Sally@pistols.firearms.com
```

After entering to whom you want to send an e-mail message (either in the form of an alias or an e-mail address), you can specify a file to be attached to this message in the optional *Attachment* field. Then you will be prompted to enter a short description of your message, called the *Subject*. This is your opportunity to grab the attention of your recipient; it is especially critical if the person receiving your e-mail gets a lot of e-mail. If the *Subject* line is empty, uninteresting, or cryptic, the addressee may not bother reading your e-mail right away, or even at all. On the other hand, a *Subject* line that is too long may be truncated, or may have its own *scroll bar*. Including a *Subject* line that is

FIGURE 1.3 Typical E-Mail Template for Message Composition.

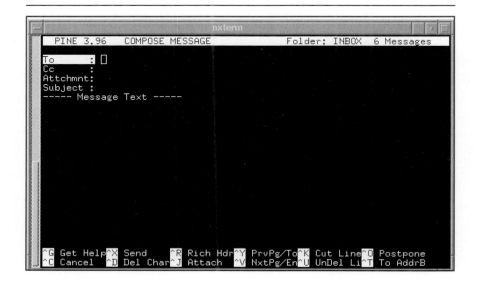

concise and descriptive is a good idea, since this introduction, along with your e-mail address, is usually the only information displayed when the recipient checks their mailbox and decides what to look at. For example, a mailer might display the information shown in Figure 1.4 when you open your mailbox.

The first message contains no subject. If you did not recognize the sender of this message, you might delete it without reading further. The subjects of the remaining messages are:

```
heads up
R: your mail
Back in the good ole USA
office hours
Re: copy edits
```

You can surmise that the message from "plum" is a warning about something. Usually, it is a good idea to read such a message right away. The message from Katharine indicates a return to the United States from a recent trip. This is an example of a good choice of *Subject*. The third and sixth messages are *replies* to messages the owner of this account sent. In one case, this is indicated with R:, and in the other case by Re:.

Do not get into the habit of just hitting the "reply" button to answer a message. After several iterations, the *Subject* deteriorates to the point where

FIGURE 1.4 Sample of a Mailer Display for Multiple E-Mail Messages.

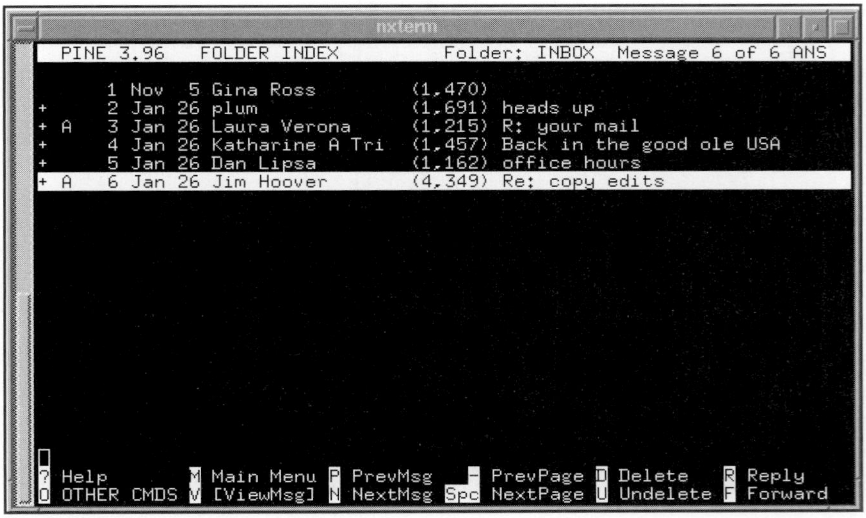

it has absolutely nothing to do with the contents of the message. Also some mailers have two reply buttons, one that lets you reply just to the sender of the message and the other that lets you reply to everyone who received the first message. Be careful, as people have embarrassed themselves and irritated other users by accidentally replying to everyone. (Most computer users hate this worse than tailgating!)

The date and time that the e-mail is sent, as well as your e-mail address, will automatically be filled in by the mailer when you send your message. You will be given the option of sending copies of the message to others when the mailer prompts you with *Cc* and perhaps with *Bcc*. Simply enter the other e-mail addresses or aliases, as desired. For important correspondence that you want to maintain a record of, it is a good idea either to *Cc* or *Bcc* yourself, unless your mailer automatically keeps sent messages.

Most mailers let you specify your favorite text editor for message composition. That is, if you are composing your message in a separate window, you can usually select the editor you would like to use. If not, you will be provided with a default editor. The text editor acts as a word processor; it allows you to type and modify your message. Also, most mailers provide on-line documentation about how to use their editor. For example, the Pine mailer uses the *Pico* (*Pi*ne *co*mposer) *text editor* by default, and documentation is available on-line to learn about both the mailer and the editor. In fact, most mail clients have an associated on-line *help* that is worth perusing. This is particularly important if you want to customize your mail client.

1.5.2   Netiquette

For the most part, when writing your e-mail message, you should follow the rules of informal letter composition. The greeting you select will often set the tone for the message. For example,

```
Dear Professor Jones,
```

is clearly very different than

```
Hey Jonesy,
```

If the person you are writing to is a close friend, you would naturally be less formal than if you are mailing your résumé to a prospective employer.

The overall tone of the message body is also very important. E-mail messages seem to be inherently direct, so it can be easy to misinterpret, or to phrase a message incorrectly. When we communicate in person (or, to a certain extent, even via the telephone), facial expressions, volume and tone of voice, and hand gestures all provide clues as to how you and the other person are reacting to the conversation. With written words, all these indicators are absent.

Informal rules of network etiquette, or *netiquette*, suggest practicing restraint when using e-mail to express opinions or ideas, especially when the message will be read by people who do not know you well. When the message is informal, a common practice is to use a *smiley* : -) or a *wink* ; -) to indicate something said in jest. These little symbols and others like them are called *emoticons* and resemble little sideways faces.

Typing a message in capital letters is considered "shouting" and doing so signals that the sender is either an e-mail novice, very angry, excited, or ignorant of the rules of netiquette. Not following the rules of netiquette (or sometimes just message content) may result in a *flaming* by someone who took offense to what you said. A flame is a nasty response from the offended party. Flaming often happens on mailing lists when one user does not show consideration for others on the list.

Figure 1.5 shows a screen shot of a graphical mail client. You should have little trouble interpreting the interface.

1.5.3 **Composition**

For sending e-mail to friends or people you know, simply type in a message as you would say it. For people you do not know, or with whom you have had little conversation, be slightly more formal and proofread your message. When applying for jobs or communicating with people for the first time, proofread and spell check your message. Many mail applications have a built-

FIGURE 1.5 Graphical Mail Client.

Netscape Mail & Discussions: Why I Run		

File   Edit   Go   Message   Communicator   Help

Get Msg   New Msg   Reply   Forward   File   Next   Print   Security   Delete

Inbox ▾ | on local machine.   0 Unread Messages, 12 Total

Subject	Sender	Date	Priority
Regarding SNER 5–year buckle	Greg Soderlund	10/08/97 19:41	
Le Grizz	Christian W Lustic	10/08/97 20:05	
Excess salt loss: reply	Bob Slate	10/08/97 20:20	
Why I Run	LeRoy Kessler	10/08/97 22:30	
Tutorial in Cambridge on Agent Pro	Eugene S Frowley	10/08/97 22:40	

▾ Why I Run

I am therefore I run.

I run, therefore I am.

It's there. I'm there.

in spell checker. A message littered with typos may offend and will certainly distract.

When sending to a large group of people, do all of the above and then reread the message one or two more times, making sure that you have phrased things in an appropriate way. Most people find that it takes much more concentration to do a thorough job of proofreading on a screen than on a printed page.

You should always "sign" your name (that is, identify yourself with your actual name or nickname, not just your e-mail address or userid) or end the message with your *signature file*. If you opt to use a signature file, many mailers will automatically append it to all messages you send. The file should contain standard contact information. For example, it might consist of your nickname, name, phone number, fax number, e-mail address (even though it's already included by the mailer), favorite quote, favorite ASCII graphic,[5] and World Wide Web address. For example, we showed the following signature in Figure 1.2:

```
                          --Alex


***************************************************************
     Alex T. Diaz         |  office:     (401) 437-2134
     332 Toast Lane       |  messages:  (401) 437-0012
     East Providence,     |  fax:        (401) 437-2137
     Rhode Island 02915   |  alex@eng.ephs.edu
***************************************************************
```

Try to limit the size of your signature file—too much can be annoying. Also, if you have frequent correspondence with a person, do not bother to include your signature file with every message. With 50,000,000 users sending 10 e-mail messages per day, each having a signature file of 200 bytes, this results in an "extra" terabyte (10^{12} bytes) of information being sent over the network every 10 days! Unless you are specifically sending someone your address or sending e-mail to someone for the first time, simply sign your name.

After composing your message using the text editor and signing it, you are ready to send. Again, for important correspondence or correspondence you would like to keep a record of, it is a good idea to *Cc* yourself on the message. Depending on the mailer, you may have to click on a "send" icon, press CONTROL-D, or press some other key or combination of keys to indicate that you want the e-mail sent. Before sending the mail, you sometimes have the opportunity (again, depending on the mailer) to *attach files* that you would like to append to the message. Some mailers also let you *insert a file* or a graphic

---

[5] ASCII stands for American Standard Code for Information Interchange. Think of it as representing standard keyboard characters. Some people consider sending an ASCII graphic poor netiquette.

at any point in the message body. Up until the time you actually send the mail, you can make changes and even decide not to send. Once the message is sent, you cannot get it back. It is like dropping a letter in a mailbox.

EXERCISES 1.5 Message Composition

22. Can your mailer handle a subject line of 500 characters in length? If not, what is the limit your mailer can handle? Write down the name of your mailer, this value, and a description of the experiments you ran.

23. Can you set the editor for message composition, or are you stuck with the default editor with your mailer?

24. Read the on-line documentation about your mail client. Report two interesting facts that you learned.

25. Compose a sample cover letter for a job to which you are applying, e-mail it to yourself, and print out the message.

26. Compile a list of your ten favorite emoticons.

27. Compose a personal signature file, send yourself a message with it appended, and print the message.

28. Print out your favorite two pieces of *ASCII art* (i.e., graphics created using just the symbols appearing on a standard keyboard) that were found in signature files.

29. Can you invoke a spell checker from inside your mailer? Write a paragraph describing how you go about doing this.

1.6 Mailer Features

The best way to learn how to use a mailer effectively is to experiment with it. This usually means sending lots of test messages to yourself. You should also read the on-line documentation, as you will usually be able to glean some knowledge about one or two lesser-known features. In this section, we mention a few features that are common to most mailers.

Most mailers provide functionality for manipulating your mailbox contents, composing messages, and saving messages to disk. For example, many mail applications allow easy access to directories or folders to organize your e-mail according to sender, subject, and so on. We will describe a generic mailer that has a *graphical user interface* (GUI). For those using a keyboard interface, the principles are similar.

A typical mailer opened in a window will contain a series of buttons (or menu items) with names such as **Compose, Copy, Delete, Edit, File, Forward,**

Move, **Next**, **Reply**, **View**, and so on. We will examine a few of these options in more detail.

1.6.1   Compose, File, and Reply

A **Compose** button typically provides the following features:

- *New*—Compose a message from scratch.
- *Reply*—Reply to the current message.
- *Forward*—Pass the message on.
- *Vacation*—You are going away and want automatic responses to be generated and to have e-mail saved.

A **File** button often has the following functionality:

- *Save*—Save the current message into a file on disk.
- *Insert*—Include a file in the body of the message being composed.
- *Exit*—Leave the mailer.
- *Open*—Open a file from disk.
- *Attach*—Append a file to a message.

A **Reply** button usually consists of the following items:

- To sender.
- To all.
- Forward (the functionality of this option is the same as that of the **Compose** button).
- Include.
- Include bracketed.

1.6.2   Bracketed Text and Include

When replying to a message, keep in mind that a period of time may have elapsed since you received the message. Thus, a reply of Yes to a message may have no meaning to the recipient. They may not recall whether you are answering the question "What is your favorite rock group?," "Do you know the French word for love?," or "Did you ace the test?"

Generally, it is a good idea to include the context of the original question along with your reply. Most mailers allow you to do this, and the format may look something like this:

```
(original text from sender)
```

if you select the include option, or

```
> (original text from sender)
```

if you select the bracketed option. "Greater than" signs (>) are usually inserted at the beginning of each line. If possible (and it generally is, since you can edit the message body), only include the original text that is pertinent to your response; do not include the whole message, unless it is brief. For example, suppose you received the message body shown in Figure 1.6. You might use the reply with the bracketed include option, as shown in Figure 1.7. If the include option is not available, be sure to give some background about the message you are responding to when replying.

We emphasize that the key to becoming competent with your mailer is to explore different options and read the help file. New mailers frequently appear, or at least new versions of old mailers are released. The features we have described may not all be present in your mailer, or perhaps they have many more capabilities. The point here is to understand the basic functionality of most mailers so you know what is out there, how to use it, and whether or not you should switch mailers.

FIGURE 1.6 Sample E-Mail Message Body.

```
Hi,

   I was wrong.  I really would like to go to dinner next
Friday.

   Oh, and by the way I got the flat tire on my bike fixed.
It was glass, ugh.

   Hope you're having a great day!

   Hasta la vista.

                        --Carme
```

FIGURE 1.7 Sample E-Mail Reply Using a Bracketed Include.

```
Dear Carme,

    >   I was wrong.  I really would like to go to dinner next
        Friday.

        Sorry.  I have made other plans.  Maybe the following
        Friday?

    >   Hope you're having a great day!

        I'm stuck at the office.  :-(

        *Tuesday*
```

1.6.3   Forwarding

At some point, you may have more than one e-mail address. For example, you might have several different computer accounts. Instead of reading mail from two different accounts, it is often more convenient to have all e-mail directed to only one account. This is usually possible by forwarding all your e-mail from one account to the other, or, in general, directing a number of e-mail addresses to one. On some systems, there is a special hidden file called something like .forward, where you can specify the e-mail address to which you would like the mail from that account to be forwarded.

Care must be taken to avoid infinite e-mail loops. For example, do not forward e-mail from account A to account B and also forward e-mail from account B to account A.

If you are going to be working at a different location for an extended period of time, it can be very useful to set up mail forwarding. This way your correspondents can continue to send you e-mail as before, even if they are not aware you are working at a new location.

EXERCISES 1.6 Mailer Features

30. What are the primary features of your mailer?

31. Explain the difference between attaching a file and inserting a file in relation to an e-mail message.

32. Create a sample e-mail message. Illustrate how the bracketed include feature can be used effectively to provide content for a reply.

33. How does your mailer distinguish between replying only to the sender versus replying to all parties receiving the message?

34. Can you "retrieve" a sent message (before it arrives at its destination) using your mailer?

35. Once you have deleted an e-mail message, does your mailer let you undelete? At what point is the message really lost?

36. Does your mailer allow you to cut and paste between different windows? That is, can you grab on-screen text and then insert it into an e-mail message you are composing?

37. Describe the process of saving an e-mail message to disk using your mailer. Does your mailer automatically save sent messages? If so, how often do you need to delete them so your disk quota is not approached?

1.7 E-Mail Inner Workings

Here we provide a simplified description of how e-mail actually works. Our goal is not to describe the low-level implementation details of a program, such as *sendmail*, but simply to acquaint you with some of the issues involved with the workings of e-mail. We will allude to postal mail for comparison purposes. In fact, we begin by considering the process you go through in mailing a piece of s-mail.

If we split the mailing of a letter into three phases, in phase one, the steps you need to perform are compose, address, package, stamp, and deposit the letter in a suitable place for pickup. In phase two, a mail person, on a fixed pickup schedule, retrieves the letter from its place of deposit. Then the letter, if correctly addressed and with the proper postage, is routed to its final destination mailbox. In phase three, the recipient checks for mail, retrieves the letter from the mailbox, opens the envelope, reads the mail, and perhaps files it away. Similar phases need to be carried out in the electronic setting.

1.7.1 Mailer, Mail Server, and Mailbox

Although the phases in the on-line world are not identical, it is helpful to use the familiar physical setting for analogies. Three main components (*mailer*, *mail server*, and *mailbox*) are necessary for the e-mail system to work. In reality, the system is much more complex.

Mailers   We introduced you to mailers in Sections 1.1 and 1.5. A mailer is also called a *mail program*, *mail application*, or *mail client*. A mailer is the software that allows you to manage, read, and compose e-mail. Think of the mailer as the function or system that allows you to perform the tasks corresponding to phase one in the s-mail analogy.

Mail servers   The *mail server* is a computer whose function[6] is to receive, store, and deliver e-mail. Conceptually, the mail server is always "listening" for the arrival of new e-mail. If new e-mail has arrived for you, the server keeps track of it. The mailer may be on the same computer that acts as a mail server.

Mailboxes   An electronic *mailbox* is a disk file specifically formatted to hold e-mail messages and information about them. Your mailbox is generally created for you by a systems administrator when you first establish your account. It is a good idea not to delete your mailbox. On some systems, even if you do remove your mailbox, it will automatically be regenerated for you when new e-mail arrives. Properly managed, the mailbox is private and only the "owner" can read from it, while everyone else can only send e-mail to it. Your mailbox is uniquely identified by your account name. Think of an electronic mailbox as the system that serves the same purpose as a mailbox in the physical setting.

There are several different ways in which users typically obtain their e-mail. We will examine two that are widely used.

1.7.2 Store and Forward Features

A mail server needs to be running nearly all the time, waiting for e-mail messages and routing them appropriately. If a mail server crashes or is down for an extended period (3–4 days), e-mail can be lost. Since a PC with an Internet connection is generally not turned on all the time, many PC users do not run mail servers locally on their computers. E-mail is very important to people, and it is unacceptable for messages to be delayed or discarded. Thus, the mail server must be a *7 by 24 machine*, that is, a machine running 7 days a week, 24 hours a day.

It is common for PC users to have their *inboxes* on a very reliable computer on which the mail server is always running. Here, we use the term "inbox" to indicate where new e-mail is stored. For some users, the inbox is the same as the mailbox. For others, the mailbox really consists of a disjointed set of file folders.

When e-mail arrives, it is saved for the addressee in their inbox until they "pick it up" by downloading the messages. The save and pick-up processes comprise the store-and-forward function.

Note that there may be a space limitation on the size of your mailbox. Generally, once this limit is reached, new incoming messages are refused until you free up space by deleting some messages. Check with your systems administrator or service provider to obtain that limit.

---

[6]   The mail server computer may have many other functions as well. It does not have to be, and usually is not, a "dedicated" mail server.

How do users obtain their e-mail? They rely on *Post Office Protocol* (*POP*) to retrieve their e-mail from a remote location. A *protocol* is a set of rules that computers use for communicating with one another. Figure 1.8 shows how the store-and-forward e-mail process works.

Let us suppose that Ricardo is sending Jane an e-mail message. Ricardo composes the message on his mail client and then selects the Send option. The message is routed via the *Simple Mail Transfer Protocol* (*SMTP*) to Jane's mail server. We have labeled the network connection between Ricardo's workstation and Jane's mail server as an SMTP link because SMTP is used to transport Ricardo's message to the mail server. Once the message arrives, the mail server stores it on disk in an area designated for Jane. The disk storage area on a mail server is often called the *mail spool*, which is the "store" part of "store and forward."

For the sake of discussion, let us suppose that

- Jane was already logged into her PC when Ricardo sent his message.

- Jane is running a POP client.

- Jane is running her mailer.

FIGURE 1.8 Example of How Users Communicate Via E-Mail in a Store-and-Forward E-Mail System.

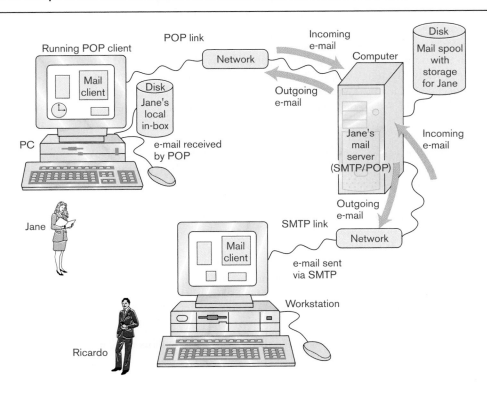

Jane's POP client knows how to communicate with Jane's mail server. Periodically, it *polls* the POP server to check if any new e-mail has arrived for Jane. In this case, there is a new message from Ricardo. His e-mail is forwarded over the network to Jane's PC and stored on her local disk. This is the "forward" in "store and forward."

We have labeled the network connection between Jane's PC and her mail server as a POP link, since POP was used to bring Ricardo's message over. The message is stored on Jane's local disk in her inbox. Jane's mailer will notify her that she has new e-mail. For example, the notification may be done via sound (with a beep) or via an icon (where a mailbox flag pops up). At this point, Jane can read and process the message. Note that if Jane were not on a PC, but were instead logged directly into her account on the mail server, there would be no POP client and dialog.

1.7.3 Central Mail Spool and IMAP

Another popular method by which users obtain their e-mail is called a *central mail spool* system as shown in Figure 1.9. This type of setup is particularly useful if someone is going to be accessing e-mail from multiple computers. In the figure, we see that Ricardo has a workstation at his office, as well as a PC at home. His inbox is maintained on a central mail server. Also, Ricardo shares his e-mail account with his son Joaquim.

Imagine what could happen if Ricardo were reading e-mail at work while his son Joaquim was simultaneously trying to read the same inbox from home. Suppose Joaquim deleted an e-mail message from his teacher addressed to his father. Would Ricardo still have access to it? If Ricardo read a message, would it be marked as "read" for Joaquim too? Lots of complications can arise in this scenario, and a protocol has been designed to handle many of the relevant issues. It is called *Interactive Mail Access Protocol* or *IMAP*.[7]

In Figure 1.9 we labeled the network connection between Ricardo's workstation and his mail server as an "IMAP link," since IMAP is used to transport messages over this connection. The network connection between Ricardo's PC and his mail server has been similarly labeled. Ricardo's e-mail remains on his mail server. The e-mail is not brought over to the computer from which he is working. Leaving the e-mail in a central location allows Ricardo to access it from several places. If the e-mail were forwarded to his workstation, there would be no way to read it from his PC, and vice versa. Various commands are available to Ricardo for manipulating his e-mail at the central site.

One advantage of IMAP over some other protocols is that it "encrypts" passwords. For example, when Ricardo is on his workstation and he wants to read his e-mail, he must send a password to the mail server; that is, he needs to authenticate himself to the mail server. Otherwise, someone else could request to read his e-mail. IMAP encodes the password so someone *sniffing* the network cannot directly obtain his password. This provides higher security

---

[7] We have also seen IMAP expanded as Internet Message Access Protocol.

FIGURE 1.9 Example of a Central Mail Spool System

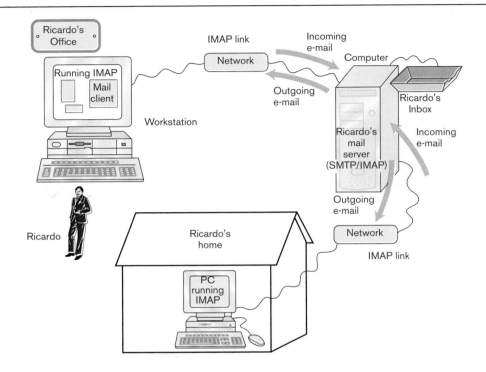

than some systems that transmit the password as "plaintext," which means that the password is not encoded. A disadvantage of IMAP or any other system is that if you spend a lot of time on-line connected to the server while reading and composing messages, this time counts against your monthly quota (if any) of connect hours. In such a situation, it may be important to learn techniques for off-line mail manipulation.

A *kilobyte* consists of 1,000 bytes and is denoted by the letter K. A *megabyte* consists of 1,000,000 bytes and is usually shortened to M. So, 1,000K is 1M. Since most servers will restrict you to 2–5M of disk space for your e-mail, and since it is common for some people to get 200K of e-mail per day, with IMAP and other similar systems, you have to be very attentive to your quota.

Other methods by which e-mail is obtained are not explored here. In addition, hybrid systems are evolving that include the best features of the currently existing systems. Depending on the application, one setup may be preferable to another. It is clear that e-mail is still rapidly changing.

1.7.4 Bounce Feature

When you send an e-mail message, the mailer software sends a copy of it over the Internet. The message has to be split up into small pieces called *packets* containing appropriate header information and *sequence numbers*. The sequence numbers are needed so the message can be reassembled in the correct order. The mailer uses the destination e-mail address to identify the computer to which the message should be routed. Eventually, the message arrives at the recipient's inbox. Some *handshaking* is needed to make sure the delivery process works smoothly. The receiving end must notify the server that all went according to plan and that the e-mail was delivered properly. This is necessary because e-mail sometimes *bounces*; that is, the e-mail is undeliverable. The major reasons for a bounce are:

1. *Bad user account name*—The e-mail bounces after it gets to the target system and that system discovers the address does not exist.

2. *Bad domain name*—This causes an immediate bounce.

3. *Domain name server is down for a number of days*—If a mail server is not working, the mail system will keep trying to send the message for a period of time. Eventually, the mail system will time out on retries, and the e-mail will bounce back.

4. *Some other malfunction*—For example, if the message being mailed is too big, a warning may be sent to notify you of this. Sometimes it is okay to ignore the warning, but other times problems might occur. For example, the message might be divided up into many small pieces that the recipient needs to assemble.

The receiving end might also need to ask the server to resend some packets, if there was a problem. If everything went well (as it usually does), your e-mail message will be waiting in your friend's mailbox when it is next checked.

E-mail capability is a feature supplied by most commercial on-line service providers, such as *America Online*. Other companies also provide e-mail only accounts, sometimes for as little as a couple of dollars per month. Naturally, the best way to locate such companies is on-line.

EXERCISES 1.7 E-Mail Inner Workings

38. Look into the *traceroute* program. What is the purpose of the program? If possible, print out a session in which you demonstrate this program.

39. Do some in-depth research about mail servers and write a short report summarizing them.

40. Which companies in your area are reputable ISPs? Can you obtain an e-mail only account? For how much per month?

41. Describe how your e-mail account works. Is it "store and forward," a central spool system, or some other type?

1.8 E-Mail Management

You have already seen that e-mail is a complex communication mechanism with many uses. Here we share a few tips that may be helpful to you. Clearly, you will develop your own e-mail style, but you should not just let it evolve without thought. It is worth spending some time evaluating how you use e-mail and how effective your responses are. Here are some questions to think about.

- Does e-mail help you at work?

- Is e-mail a waste of your time?

- Are you flooded with e-mail from mailing lists?

- Is e-mail a distraction?

- Are you constantly reading forwarded jokes?

- How can e-mail make your life more enjoyable?

- How can e-mail make you more productive instead of less efficient?

- Do you receive a lot of useless gossip?

A new e-mail arrival is usually signaled to you by your mailer. If you are already logged on, there may be a beep, which can usually be suppressed if you find it annoying, or perhaps something like an icon of a mailbox with a flag up. If you are just logging on, a note may be printed on the screen that says you have new mail.

When you decide to view your e-mail, your mailer will provide some sort of index of messages, with the subject line displayed (as in Figure 1.4). Usually, the messages are numbered in sequence. They might be displayed in either chronological or reverse chronological order. The mailer typically displays the first or current message.

1.8.1 Action Options

At this point, you have a number of options for dealing with the message. A few are listed here:

1. You might decide, based on the subject line and the address of the sender, that you want to delete the message without reading it. This is one way to deal with junk e-mail and one reason to make sure that your subject line makes sense when sending mail.

2. You may decide that you do not have time to read the message right now and that you will get back to it later. In this case, you could simply skip over it or save it to a file.

3. You may decide to read the current message now. After reading the message, you have the option(s) of deleting the message, replying to it, forwarding the message to someone else, saving the message in a file, or saving the message in the mailbox. Note that it is sometimes worth scanning your entire mailbox before replying to a message, since another later message may supersede the contents of the earlier one.

If you do not receive a lot of e-mail (say, less than twenty messages per day), it may be tempting to let them "lie around" in your mailbox. However, if the volume of e-mail you receive picks up, either because you find that you really like this method of communicating or because you subscribe to one or more mailing lists, you will need another strategy for dealing with your e-mail.

One recommended strategy, called *triage*, can be summed up as follows:

1. Skim for the most important messages (from your best friend, boss, and so on).

2. Skim for what you can delete unread.

3. Then work through the remainder.

Another possible strategy, called *skim and delete*, works as follows:

1. Skim through your mailbox, reading only those messages that are important to you while deleting the rest.

2. If possible, deal with each message immediately and generate a response, if necessary.

3. If the message requires more than a couple of minutes to address, save it for later, if time does not permit handling it now.

If a message is very important, you should save it. Messages can be stored in folders organized by subject, date, and so on, or they can be saved in your mailbox.

Naturally, the mileage you get from such strategies will vary. However, it is critical to develop some sort of protocol for dealing with e-mail, especially if you find it becoming a burden.

1.8.2   Vacation Programs

If you receive a lot of e-mail, you may consider the possibility of configuring a *vacation program* when you go away for an extended period of time. A vacation program is one that automatically replies to your e-mail. Usually, the program sends a brief reply back to each message you receive. For business purposes, it is customary to include the name and telephone number/e-mail address of someone to contact in your absence. You should be aware that a large number of users despise vacation programs. Consider the following points before routinely setting up such a program.

1. Do most of your friends know you are going away?

2. Do most of your business associates realize you are on vacation?

3. Are you subscribed to any mailing lists where 1,000 or so innocent users could be bombarded by your vacation program?

4. If someone knows you are away for a week, will it make a big difference to them?

5. Do you want people to know you are away, especially strangers? What if they use the Internet to figure out where you live? Or, what if they decide that while you are away, it would be a good time to attempt to break in to your account, since it will probably go unnoticed for a while.

6. Do you want to generate lots of additional and perhaps unnecessary e-mail?

7. Does your "vacation" message tell recipients who to contact in your absence?

Not all vacation programs are created equal. With a good one and the right mailing list server software, things can work very well. Nevertheless, at least think about the points mentioned here before installing such a program.

1.8.3 E-Mail and Businesses

When working in a business environment that uses e-mail, you should be aware that it is currently legal for an employer to read all company e-mail. Very few companies actually do read employees' e-mail, but you should be aware that they can. A company could maintain backups of all e-mail for a long period of time. If necessary they could go back and review the e-mail messages of an employee. Such backups can also be subpoenaed.

Businesses sometimes use *e-mail filters*. The filters can work in both directions, to limit either incoming or outgoing e-mail. The filtering mechanism examines each message's e-mail address before deciding whether or not to send the mail on. Businesses use e-mail filters to restrict with whom their employees can communicate.

EXERCISES 1.8 E-Mail Management

42. For a one-week period, chart the number of messages you receive per day, how many you read, and how many to which you actually reply. What percentage of the messages is useful?

43. Write a short summary describing your most embarrassing or frustrating e-mail experience.

44. What have you found to be the most helpful strategy for managing your e-mail?

45. How do you set up a vacation program for e-mail on your computer system?

46. Estimate how much money you save per month by using e-mail instead of the telephone.

1.9 MIME Types

As previously stated, MIME is an acronym for *Multipurpose Internet Mail Extensions*. Originally, only plain ASCII text files could be sent via e-mail. Today, an e-mail message may contain an attachment that consists of virtually any type of file. Usually, people refer to ASCII files as text or *plaintext* files and to all other types of files as *binary files*. For example, in addition to text, another form of media such as graphics, *HTML* code (see Chapter 2), a spreadsheet document, video, voice, and/or a word processor document could also be attached to a message. All that is necessary is that your mailer and the recipient's mailer be MIME compliant.

For example, suppose you want to send a word processor document that has a group of tabular columns with complex formatting. If you try to transfer the file as text, all the formatting may be lost and the message will likely appear jumbled, if at all. However, having MIME-compliant mailers at both ends takes care of the messy details and the message arrives intact, as desired. If the recipient has the corresponding application, they will then be able to view the document.

Suppose someone sends you a message that contains an attachment consisting of an HTML document. If your mailer has a graphical user interface and is MIME compliant, it may display an icon indicating that the file attachment is an HTML document. In some mailers, if you select the attachment for viewing, an HTML previewer is automatically launched that renders the HTML document for you. You do not have to save the document and then run a program to preview the document; the mailer takes care of everything for you. This is very convenient.

We should point out that some security problems exist with MIME. For example, suppose you receive an e-mail message that has an attached Microsoft Word file. Assuming you have configured your mailer accordingly, when you select the attachment to view, your mailer will open up the Word document for you. If a clever, malicious user sent you the message, they may have included one of the many *Microsoft Word Macro Viruses*. At this point, it is possible that the intruding program can *infect* your files. Because of this and other similar security problems, some locations are reluctant to support MIME. To avoid

TABLE 1.3 Common MIME Types.

Type	Subtype	Description of Content Type	File Extension(s)
Application	postscript	printable postscript document	`.eps, .ps`
	tex	TEX document	`.tex`
	troff	printable troff document	`.t, .tr, .roff`
Audio	aiff	Apple sound	`.aif, .aiff, .aifc`
	au	Sun Microsystems sound	`.au, .snd`
	midi	*Musical Instrument Digital Interface*	`.midi, .mid`
	realaudio	Progressive Networks sound	`.ra, .ram`
	wav	Microsoft sound	`.wav`
Image	gif	*Graphics Interchange Format*	`.gif`
	jpeg	*Joint Photographic Experts Group*	`.jpeg, .jpg, .jpe`
	png	*Portable Network Graphics*	`.png`
	tiff	*Tagged Image File Format*	`.tiff, .tif`
Model	vrml	*Virtual Reality Modeling Language*	`.wrl`
Text	html	*HyperText Markup Language*	`.html, .htm`
	plain	unformatted text	`.txt`
	sgml	*Standard Generalized Markup Language*	`.sgml`
Video	avi	Microsoft *Audio Video Interleaved*	`.avi`
	mpeg	*Moving Picture Experts Group*	`.mpeg, .mpg`
	quicktime	Apple QuickTime movie	`.qt, .mov`
	sgi-movie	Silicon Graphics movie	`.movie`

this problem, you can have your mailer save attachments to disk and then run a *virus check* on them before opening the files. We should mention that, to date, there is no way for your computer to be infected with a virus by the simple act of reading a non-MIME message.

A number of MIME types are important and will play a significant role in later parts of this book. In Table 1.3 we summarize a few of them.

EXERCISES 1.9 MIME Types

47. What happens if your mailer supports MIME, but the recipient's mailer does not?

48. Investigate how MIME works and write a summary of your findings. Try to track down a *Request for Comments* (*RFCs*) about MIME.

49. Does your e-mail client provide you with any security mechanisms, that is, can you encrypt a message? Explain.

Jump Start: Browsing and Publishing

2

2.1 Introduction

This chapter provides the background necessary to start using the World Wide Web and to create Web pages. We will introduce the following topics:

OBJECTIVES

- Web browsers.
- Web surfing.
- *HyperText Markup Language* (HTML).

- Web page installation.
- Web page setup.
- HTML formatting and hyperlink creation.

2.2 Browser Bare Bones

We begin with browser essentials to get you up and surfing the Web. Even if you have already been browsing the Web, please read on; a few pointers may be useful.

Very few agreed-upon precise definitions have been given for new terms involving the Internet, such as *Web browser*. A Web browser is one of many software applications that function as the interface between a user and the Internet. The browser not only sends messages to Web servers to retrieve your page requests, but also *renders* the HTML code once it arrives. That is, the browser interprets the code and displays the results on the screen. Many browsers have built-in mail clients and/or newsreaders. Additionally, auxiliary programs such as *helper applications* and *plug-ins* can be configured into the browser. (We will describe these features in more detail later.) It is safe to assume that browsers will continue to grow in complexity and functionality in the foreseeable future.

Popular browsers include Netscape Navigator, Microsoft's Internet Explorer, Mosaic, and Lynx. The first few are *graphical-based Web browsers*, whereas Lynx is a *text-only browser*. This makes Lynx very fast, since graphics often cause Web pages to load more slowly. For our presentation, we will describe the Netscape Navigator, although most browsers have similar features (the exception being Lynx).

2.2.1 Browser Window Terminology

Figure 2.1 illustrates a sample browser window. The different components of the window are numbered. We will provide a name and a short explanation of each part and will then discuss the parts in more detail.

1. *Title bar*—The location where the document's title is displayed.

2. *Menu bar*—The place showing the headings of the main pull-down command menus.

3. *Toolbar*—The area providing access to a number of single-mouse-click commands.

4. *Location*—The area where the *Uniform Resource Locator* (URL) (discussed in Section 2.3.2) of the document is displayed. The item in the location field usually begins with `http://`, although you may also see `file://`, `ftp://`, `gopher://`, `javascript:`, `mailto:`, `news:`, or `telnet://`.

5. *Hot buttons*—Single-click buttons that provide a number of convenient features.

6. *Netscape icon*—An image that shows movement to indicate when a document is being downloaded from the Internet.

7. *Scroll bar*—Arrows that allow the user to display a different part of a "large" document.

8. *Document area*—The part of the window that is used for displaying the currently loaded document.

9. *Status bar*—A field used to convey helpful (and current) information to the user, such as a URL or a programmer-specified message.

10. *In-line image*—An image appearing within a document.

11. *Hyperlink*—A highlighted (usually underlined) part of a document that, when selected, causes the browser to retrieve and display a (new) document.

FIGURE 2.1 Sample Screen Illustrating the Terminology Associated with a Browser Window.

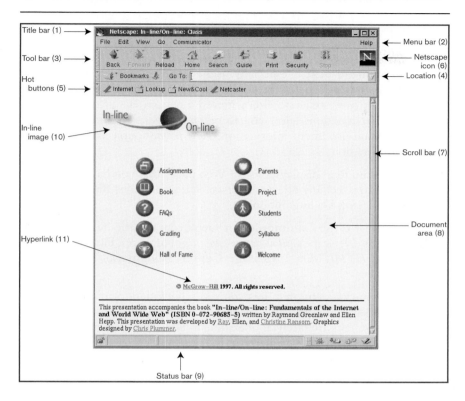

2.2.2 Menu Bar

Looking at the Netscape[1] window in Figure 2.1, you will notice the items **File**, **Edit**, **View**, **Go**, **Communicator**, and **Help** in the menu bar. (You will see different items, such as **Bookmarks**, **Options**, **Directory**, and **Window**, in Netscape 3.x.) As is customary when using menus, menu items that are not available to the current configuration of the browser are shown in a lighter color gray ("grayed out").

- The **File** menu item will allow you to launch a new browser, utilize the Netscape mailer, open a new URL, open a local file, save a file, print a screen, or exit the browser. The **Send Link** option provides a convenient

---

[1] It is common to refer to the Netscape Navigator by the name of the company, Netscape, that developed it. Netscape's terminology is very confusing. With version 4.x of their browser, it is still the Navigator, part of a suite called *Communicator*, but many people now call the browser Communicator.

way to e-mail the URL of the page you are currently visiting to someone else. If you are surfing the Web from someone else's account, this is a good way to send yourself a URL so you can bookmark it later.

- The **Edit** button provides basic text editing capabilities.

- The **View** menu item is especially useful because it allows you to view the HTML source code of the document being displayed. This is a great way to learn how someone else achieved a certain layout on their Web page. Other functions allow you to reload a page, load images, refresh a page, obtain document information, view *frame* source code, and view frame information.

- The **Go** menu item displays a list of Web pages that you have visited and allows you to select any one to return to. It also provides **Back**, **Forward**, **Home**, and **Stop** loading options.

- In Netscape 4.x, the **Communicator** menu contains the items **Collabra Discussions** (Netscape's *collaborative computing* software), **Page Composer** (Netscape's *HTML editor*), **Message Center**, **Bookmarks**, and **History**, among others.

- The **Bookmarks** menu item lets you add a bookmark or directly select a (previously saved) bookmark. A *bookmark* is simply a saved Web location (i.e., URL). URLs are often cumbersome to type. For Web pages you visit frequently or just want to remember, the bookmark mechanism is a handy tool.

- The **Help** menu item provides Netscape help and information.

- (Netscape 3.x) The **Options** menu item has a number of features that allow you to customize your browser. For example, you can toggle the images setting (see page 41), specify *cache* size, and allow or disallow *cookies* to be written.

- (Netscape 3.x) The **Directory** menu item provides a vertical list of the hot buttons.

- (Netscape 3.x) The **Window** menu item provides access to news, mail, your address book, bookmarks, and a history mechanism.

2.2.3 Toolbar

The toolbar is located under the title bar and contains buttons for **Back**, **Forward**, **Reload**, **Home**, **Search**, **Guide**, **Print**, **Security**, and **Stop**. (You will see different items, such as **Images**, **Open**, and **Find**, in Netscape 3.x.)

Back   A browser normally saves copies of the pages you have viewed in a *cache*. Think of a cache as local computer disk space from which the browser can quickly retrieve a document. For example, suppose you load a

Web page with lots of graphics from a Web site located across the country. If this document is cached locally, then the next time you request it, the browser can load the copy stored in cache. The document can thus be loaded much faster. How much disk space should be allocated to cache? This is an option that the user can set. What if a document is updated but you keep retrieving the old cached version? You can try to reload the document and, if necessary, clear the cache. Many other interesting issues involving caches are dealt with in introductory computer science courses.

Suppose you just visited Web pages A, B, C, D, and E, in that order. At this point, hitting the **Back** button would take you from E to D. Clicking on **Back** again takes you to C, and so on. The **Back** button allows you to access the most recently visited page without typing in its URL. The page is usually loaded quickly, since it is available from cache.

Forward   The **Forward** button allows you to page forward in much the same way that the **Back** button operates.

Reload   The most current version of a Web page can be loaded by clicking on the **Reload** button. This is particularly useful if you have just modified the source code for a page and want to view and check the changes.

Home   The user can specify what Web page to load when the browser is first activated. This page is often called the *homepage*. The **Home** button will load the homepage that has been designated. The default page is often set to the Web page of the company that developed the browser.

Search   Clicking on the **Search** button brings up one of the many useful search tools that Netscape "knows" about. Once the search tool is loaded, you can use it to search the Internet.

Guide   The **Guide** button leads you to a mini information center (provided by Netscape) from which you can locate all kinds of useful items. The information displayed is updated frequently.

Print   You can obtain a hardcopy of the currently displayed Web page by clicking the **Print** button.

Security   The **Security** button allows you to examine and specify security options.

Stop   The **Stop** button is used to stop the transfer of a Web page. This can be handy if you realize that you have selected the wrong link, or if the page you selected is going to take too much time to load. This button also allows you to stop endlessly looping animated GIFs.

Images   (Netscape 3.x) The **Images** button lets you "toggle" the state of image downloading; that is, if the browser is currently downloading images, selecting the **Images** button tells the browser not to download images. If the browser is not downloading images, then selecting the **Images** button tells the browser to download images. Since images require a lot of storage and therefore a lot of time to download, this button comes in handy when you do not need or wish to download images.

Open   (Netscape 3.x) The **Open** button provides you with a dialog box in which to type a URL. When you hit the return, the browser requests and then renders the Web page you specified.

Find   (Netscape 3.x) The **Find** button initiates a search within the current Web page for a word or phrase that you specify. If the pattern of the word or phrase occurs in the document, the browser scrolls the page to the first occurrence of that pattern and then highlights it. If the pattern is not found, the browser will usually ask if you want to search in the reverse direction. (This function is now located in the Menu bar under the **Edit** entry in Netscape 4.x.)

2.2.4   Hot Buttons

Beneath the location area are the hot buttons (also called *directory buttons*) that Netscape provides. Other browsers provide their own versions of these buttons. These buttons include:

- **Internet**   This button has the same effect as the **Guide** button of the tool bar.
- **Lookup**   This button contains two options: **People** and **Yellow Pages**.
  - **People**   Various search programs that are available to locate an individual.
  - **Yellow Pages**   Various search programs that are available to locate a business.
- **New&Cool**   This button contains two options: **What's New** and **What's Cool**.
  - **What's New**   A list of new, interesting Web pages.
  - **What's Cool**   A selected list of "cool" Web pages.
- **Netcaster**   This button takes you to information about Netscape's Netcaster product, which allows you to open a "channel" to receive a continuous flow of information to your computer.

The following buttons appeared in earlier versions of Netscape.

- **Destinations**   This allowed access to a list of "cool" hyperlinks.
- **Net Search**   This provided a quick way to access a variety of search programs.
- **Software**   This button accessed information about Netscape software that was currently available for downloading.

2.2.5   Hyperlinks

Let us elaborate on the important concept of a *hyperlink*. Hyperlinks are click-able text and/or images that generally cause the downloading and rendering of a new HTML document. Hyperlinks are often displayed in a different text color than the remainder of the document, and they are usually underlined to make them stand out. An image serving as a link may have a border around it that is the same color as other hyperlinks on the page. In either case, moving the mouse over a hyperlink (termed *mousing over* a hyperlink) will cause the mouse cursor to change appearance, perhaps from an arrow to a hand.

The location (URL) of the link being moused over will be displayed in the status line. It is very helpful if you understand URLs. An experienced user, upon seeing a URL, will know where the document is stored and approximately how long it should take to download (assuming an educated guess can be made as to the size of the document).

EXERCISES 2.2 Browser Bare Bones

1.   Print the source code for your favorite Web page. Print the browser screen of your favorite Web page and label the different components of the page on the hardcopy.

2.   Explain the types of facts that are available when viewing document informa-tion. For example, for an HTML document, can you tell how many bytes long it is or when it was created? What about the size of an image?

3.   Is there a difference between typing in a URL in the location field versus selecting the **Open Page** option of the **File** button?

4.   Experiment with the history mechanism of the browser. Write a paragraph explaining how it works.

5.   What options are available for customizing your browser? Elaborate.

6.   Skim through the on-line help for your browser. Summarize in two paragraphs what type of information is available.

7.   Using the hot buttons, try to track down your parents or a friend on the Web. Describe how you proceeded and whether or not you were successful.

2.3   Coast-to-Coast Surfing

We are now ready to start using the browser to discover information on the World Wide Web. As is customary, we have been shortening the phrase World Wide Web to "Web." Other common short forms are WWW, W3, and W^3.

The Web provides a means of accessing an enormous collection of infor-mation, including text, graphics, audio, video, movies, and so on. One of the most exciting aspects of the Web is that information can be accessed in a nonlinear and experimental fashion. Unlike reading a book by flipping to the next page in sequential order, you can "jump" from topic to topic via hyperlinks. This nonlinear approach to information gathering, or browsing, is sometimes referred to as "surfing the Web." As a reader, you have the option to select what to explore next. Different readers will proceed through the same Web presentations in totally different ways, depending on their backgrounds, needs, and personalities.

2.3.1 Web Terminology

Web surfing is a great way to become familiar with the Web. To begin our discussion of Web surfing, we first introduce and review some common Web terminology[2]:

- **Page** or **Web page**   A file that can be read over the World Wide Web.

- **Pages** or **Web pages**   The global collection of documents associated with and accessible via the World Wide Web.

- **Hyperlink**   A string of clickable text or a clickable graphic that points to another Web page or document. When the hyperlink is selected, another Web page is requested, retrieved, and rendered by the browser.

- **Hypertext**   Web pages that have hyperlinks to other pages. More generally, any text having nonlinear links to other text.

- **Browser**   A software tool used to view Web pages, read e-mail, and read newsgroups, among other things. Browsers are also called *Web clients*.

- **Multimedia**   Information in the form of graphics, audio, video, or movies. A multimedia document contains a media element other than just plaintext.

- **Hypermedia**   Media with links and navigational tools.

- **Uniform Resource Locator**   A string of characters that specify the address of a Web page.

- **Surfer**   A person who spends time exploring the World Wide Web.

- **Web presentation**   A collection of associated and hyperlinked Web pages. Usually, there is an underlying theme to the pages. For example, a Web presentation for a company may describe facts about the company, its employees, its products, and the method for ordering the products on-line.

- **Webmaster**   A person who maintains, creates, and manages a Web presentation, often for a business, organization, or university. This person

---

[2] In a number of sections in this book, we introduce special terminology. The terms are presented in a logical order, rather than an alphabetical order. The terms may be found in alphabetical order in the glossary and index.

usually "signs" Web pages, so that questions and comments can be sent to them.

- **Web manager**   Synonym for Webmaster.

- **Web site**   An entity on the Internet that publishes Web pages. A Web site typically has a computer serving Web pages, whereas a Web presentation is the actual Web pages themselves. For example, `www.lsu.edu` is the name of a Web site, whereas

```
www.lsu.edu/~holmes/index.html
```

is the name of a Web presentation.

- **Web server**   A computer that satisfies user requests for Web pages.

- **Mirror site**   A site that contains a duplicate copy of a Web presentation from another site. If a Web presentation is extremely popular, other sites may be used to mirror the original presentation; that is, they contain the same information as the original site. This allows the load on the Web server and the network to be distributed. If one server is down, a mirror site can be tried. If several mirror sites exist, it is a good idea to try the one closest to you first.

2.3.2   Uniform Resource Locator (URL)

In Section 2.2, we mentioned that the address of the Web page being displayed is shown under the toolbar in the location area of the browser window. This Web page address is a URL (pronounced "you-are-ell" or sometimes "earl"). Typing a URL in the location area and hitting the return key will cause the browser to attempt to retrieve that page. If the browser is successful in finding the page, the browser will display it. This high-level explanation does not, however, convey any of the details of what is happening. To go from a URL to having the Web page displayed, the browser needs to be able to answer such questions as:

1. How can the page be accessed?

2. Where can the page be found?

3. What is the file name corresponding to the page?

The URL is designed to incorporate sufficient information to resolve these questions. Quite naturally, then, the URL has three parts. We can view the format of a URL as follows:

```
how://where/what
```

At this point, it is helpful to consider a sample URL to illustrate the three parts:

```
http://pubpages.uminn.edu/index.html
```

Let us break this example down into its components.

1. http—Defines the *protocol* or *scheme* by which to access the page. In this case, the protocol is *HyperText Transfer Protocol*. This protocol is the set of rules by which an HTML document is transferred over the Web (see further comments about index.html).

2. pubpages.uminn.edu—Identifies the domain name of the computer where the page resides. The computer is a Web server capable of satisfying page requests. Just as a waiter serves food, a Web server "serves" Web pages. The name pubpages.uminn.edu tells the browser on which computer to find the Web page. In this case, the computer is located at the University of Minnesota.

3. index.html—Provides the local name (usually a file name) uniquely identifying the specific page. If no name is specified, the Web server where the page is located may supply a default file. On many systems, the default file is named index.html or index.htm.

This example demonstrates that the URL consists of a protocol, a Web server's domain name, and a file name.

Like a social security number (SSN), which uniquely identifies a person, URLs uniquely identify Web pages. An SSN is an identifier; it indicates where someone lived regionally when their SSN was issued, and the year of issue. For example, 001 through 003 are for New Hampshire. The middle two digits are an indirect code for the year of issue. However, based on a person's social security number you cannot tell whether they currently live in Alaska or Rhode Island; you also cannot tell what type of job they have. In comparison, the URL provides all the information a browser needs to locate and access a Web page anywhere in the world. The URL format is somewhat flexible so that the system can be adapted when necessary.

Entering a URL in the Location field of the browser will bring up the designated Web page, barring any problems. For example, if the Web page has moved to another machine or has been removed, or if you type an invalid URL, or if the server you are trying to access is unavailable, an error message will be displayed. Another way to retrieve a Web page is to mouse over and click on a hyperlink in the Web page that is currently being displayed. Recall that a hyperlink is a string of text or a graphic that points to other pages.

In the URL example presented earlier, the protocol to access the page was http. This is used for transferring an HTML document. Much of the power of browsers is that they are *multiprotocol*. That is, they can retrieve and render information from a variety of servers and sources. Table 2.1 provides a summary of other common protocols.

TABLE 2.1 Protocols That May Occur in URLs.

Protocol Name	Use	Example
ftp	File transfer	`ftp://ftp.bio.umaine.edu`
gopher	Gopher	`gopher://gopher.tc.umn.edu/11/Libraries`
http	Hypertext	`http://www.chem.uab.edu/ pauling/argon.html`
mailto	Sending e-mail	`mailto:kim-lee@mycompany.com`
news	Requesting news	`news:soc.penpals`
telnet	Remote login	`telnet://www.amnesty.org/`

EXERCISES 2.3 Coast-to-Coast Surfing

8. Surf the Web and locate three Web pages that contain glossaries of computer jargon. List three terms you were previously unaware of and their definitions.

9. Compare and contrast an e-mail address and a URL.

10. Give legal URLs for seven different top-level domains.

11. When a Web page is requested, a number of different error messages are possible. List their numerical codes and describe what each one means.

12. Can you locate any information about the `file` protocol? Describe your findings.

2.4 HyperText Markup Language: Introduction

Here we describe some basic HTML *tags* to get you started publishing a Web page, and we introduce the most useful *attributes* of each tag. Since HTML is not completely standardized yet, and since there is room for differences in interpreting and implementing a standard, it is possible that not all versions of all browsers will support all attributes.

A Web page is created when an ordinary ASCII text file is "marked up" using HTML tags and is then displayed using a browser. The tags are predefined combinations of characters enclosed between ⟨ and ⟩ characters. These symbols are called "less than" and "greater than," respectively. Sample tags are ⟨HTML⟩, ⟨CODE⟩, and ⟨TITLE⟩. The tags are embedded within the text of a file, and they indicate how the text is to be interpreted and displayed by the browser. The word "markup" is used because copy editors use similar notations for editing printed matter.

How a Web page looks when displayed depends on (at least) three things:

1. The HTML tags used.

2. The specific browser rendering the page.

3. The user's system and monitor.

HTML tags do not define exactly how the Web page is supposed to look; rather, the tags describe how the elements of the page, such as headings, lists, paragraphs, and so on, are to be used. For example, many people think of a heading as being numbered, appearing in boldface, and being displayed in a larger font. The Web browser actually formats the HTML document and different browsers may display the (same) HTML code differently. This is a point worth repeating. What you see using your browser may be significantly different than what someone else sees when viewing the same Web page. Since not all monitors support the same set of colors, the quality of the user's monitor affects the appearance of the colors in a document.[3] That is, your cranberry color may be very different than somebody else's.

HTML is not case sensitive. That is, the tag ⟨HTML⟩ means the same as ⟨html⟩, which means the same as ⟨Html⟩. However, you should be consistent with your tags, since it will make them easier to locate when you are editing files or debugging your code. Some Web authors prefer to use all capitals or a particular color for their tags, as these make the tags stand out from the remainder of the document. We follow both of these conventions throughout the book.

2.4.1 HTML Tag Syntax

Fortunately, learning HTML tag syntax is easy. The basic form for all HTML tags can be written abstractly as

```
<TAG ATTRI1 = "V1" ATTRI2 = "V2">item to be formatted</TAG>
```

where ATTRI means *attribute*. TAG means any HTML tag.

Many HTML tags have attributes. In the general form presented here, we have listed two attributes, called ATTRI1 and ATTRI2. The number of attributes varies from tag to tag. Attributes typically have a choice of several values. In the expression we gave, the values are denoted V1 and V2, respectively. Note the equals (=) sign; this is programming syntax for assigning ATTRI1 the value V1. Also notice that we put quotes around V1 and V2. For all HTML attributes, it is safe to quote their values. However, if the value is a

---

[3] Actually, it is more than just the monitor. Even if you come down to the basic 256 colors, Windows PCs and Macs, for example, use inherently different color palettes, with only 216 colors in common. Also, the brightness setting on the two kinds of systems is different.

number, it is sometimes not necessary to quote it. Nevertheless, we prefer to quote all values. Also observe that we have not left any white space between the item to be formatted and the surrounding tags. This is a good habit to get into, as otherwise the hyperlinks you format will not appear as you might want.

TAG has a corresponding *ending tag* (also referred to as a *closing tag*), namely </TAG>. The ending tag is the same as the starting tag except for the "/" character. Not every tag has an ending tag, but most do. Also, some ending tags are commonly omitted. We will mention these where appropriate. Ending tags can always be identified by the forward slash preceding the tag name. If you keep these basic rules of syntax in mind while learning new HTML tags, you will have an easier time coding properly.

Many new users of HTML begin learning to program by cutting and pasting code from existing Web pages. Although it is often helpful to look at someone else's HTML code, we recommend against copying it, for the following reasons:

1. Copyright issues must often be considered.

2. A great deal of HTML code is poorly written.

3. A lot of HTML code contains bugs.

4. It is easy to fool yourself into believing you have learned the basic elements of HTML.

5. A significant fraction of HTML documents have inconsistent styles within them, since they have been cut and pasted together many times over.

Occasionally, you may find that someone else did something so well you want to "borrow" it. In such cases, it is a good idea to ask the person for permission to use the code or to credit the person's work, as appropriate.

2.4.2 HTML Document Creation

To produce an HTML document, you need to use a text editor to create an ASCII file with an extension of .html or .htm. Remember the MIME type file extensions for HTML given in Table 1.3? The file you produce must contain correct HTML code so the browser can render it. Once you have constructed and saved the file, you need to set the file *permissions* accordingly so that other people on the Web can access the document. In Section 2.5, we describe the basics of installing a Web page, and we address such issues as file protections. Here we focus on creating a simple HTML document.

Every HTML document has two parts: a *head* and a *body*.[4] The associated HTML tags for these parts are <HEAD> with closing tag </HEAD>, and <BODY> with closing tag </BODY>. Surrounding all the text in the entire file are the

---

[4] There are a few exceptions, such as frameset pages.

beginning and ending HTML tags—<HTML> and </HTML>. These tags let the browser know that the file is indeed an HTML file. If the browser tried to render a Visual Basic program or a Word file as an HTML document, there would be lots of problems. Thus, the opening <HTML> tag and the closing </HTML> tag give the browser the go-ahead to render the file as an HTML document.

A title tag, <TITLE>, is contained within the head of the document to provide a title for the document. Its corresponding ending tag is </TITLE>. Do not include any HTML formatting within the title tag. The title should provide a concise description of the page, since the title is prominently displayed in the browser window's Title bar when the page is being viewed. Perhaps even more important is the fact that the title is also used as the default *bookmark description* when a Web page is bookmarked. A title such as "My Homepage" is a poor choice, whereas "Sung Lee's Homepage" is much more descriptive. Finally, an HTML document's title can affect how the document is indexed by some *search engines*.

To demonstrate the basic elements of an HTML document, we will create a simple one and place it in a file called cone.html, standing for "creation one." In our example, "In-line/On-line: Creation Number One" is an appropriate title. As described so far, we have the following HTML code in cone.html:

```
<HTML>

<HEAD>
<TITLE>In-line/On-line: Creation Number One</TITLE>
</HEAD>

<BODY>
</BODY>
</HTML>
```

Notice that the spacing (i.e., putting each item on a separate line) makes the code easy to read. Compare this with the following:

```
<HTML> <HEAD> <TITLE>In-line/On-line: Creation Number
    One</TITLE> </HEAD> <BODY> </BODY> </HTML>
```

We suggest you take the time to make your code readable. In the long run, you will save time editing and debugging. Some authors also use indentation to *pretty print* their code. For example:

```
<HTML>
    <HEAD>
        <TITLE>In-line/On-line: Creation Number One</TITLE>
    </HEAD>
    <BODY>
    </BODY>
</HTML>
```

Looking at `cone.html` using your browser will not be very exciting at this point since the document area of the browser window has nothing to display. The only thing you will see is the title, "In-line/On-line: Creation Number One," in the Title bar. To display some words in the document area, add text between the `<BODY>` and `</BODY>` tags, as follows:

```
<BODY>
This is my first Web page.  Yeah!  I love
electronic publishing.  Hi Mom.  Hi Dad.  Hello world!
</BODY>
```

In Figure 2.2, we show the results of viewing `cone.html` with a browser. Notice the text in the document area of the browser window. Closely observe how the text is positioned in the browser window here and in succeeding examples. Browsers incorporate a simple *line breaking algorithm* to wrap the text if it would be too wide for the window's document area. If a window of another width is used, the text will wrap differently.

EXERCISES 2.4 HyperText Markup Language: Introduction

13. Create a file called `index.html` having the basic set of tags that any HTML document should contain.

FIGURE 2.2 The File `cone.html` **as Rendered by a Browser.**

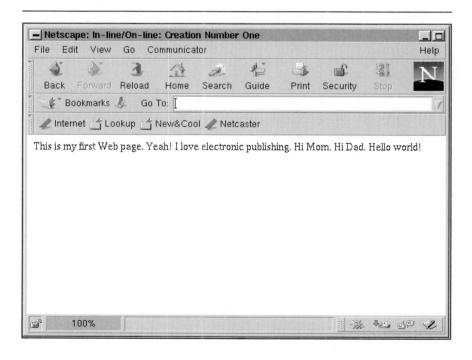

14. A field hockey player's coach asks her to design a Web page for the university's team. What are two possible titles for it? Create the code for a simple version of the page using one of your titles.

15. Create an HTML file that contains all the basic tags you have learned so far. Include a title for the page that would be a sensible title for your *personal page*. A personal page, also called a *homepage* by some people, is an individual's top-level HTML document. A personal page typically contains data about the individual, contact information, a table of contents to the Web presentation, and so on.

16. Write two paragraphs explaining to a friend, who has a limited knowledge of computers, how to prepare a basic HTML document.

17. Does your browser hyphenate words when it wraps the end of lines? Is hyphenation a choice that you can toggle on or off?

2.5 Web Page Installation

In order to view your pages on the Web, you will need to install them on a *Web server*. A Web server is a program, located on a computer with Internet access, that responds to a browser's request for a URL. That is, a Web server meets the demands of users by supplying or serving them the Web pages requested. Ideally, the server should have an uninterrupted Internet connection, so that the pages it handles are always available.

A Web server is accessible to many of us through work or school, and many ISPs include space on their computers that run the Web servers as part of the basic set of services covered in their monthly fee. The systems administrator responsible for the server can usually fill you in about the site-specific details for publishing your Web pages. In this section, we describe, at a high level, the basic steps necessary for installing your Web presentation. We then take a more detailed look at how Web pages can be set up on a UNIX-based Web server. UNIX is a type of computer operating system. Appendix D contains an introduction to UNIX. Other Web servers, such as those that are Windows based, will require a different installation procedure for Web pages.

Why are we discussing the UNIX platform?

- The basic principles we describe for UNIX can be applied to several other systems.

- The first Web servers developed were built on the UNIX platform.

- A huge number of sites are currently running UNIX-based Web servers. In July 1997, there were about 550,000 UNIX Web servers running *Apache* (the most popular Web server on the Internet).

- UNIX is prevalent in academic settings.

- Many new Web server features appear first on UNIX-based servers.

- The source code is often free for programs developed under UNIX, or at least some version of the software is often available for public use.

We should note that desktop operating systems are increasingly shipped with a simple Web server that can optionally be set up and used. This will eliminate one more distinction between server and client. The last big remaining distinction will be accessibility, that is, whether the server is normally on-line.

2.5.1 Basic Principles

What items are necessary for someone in, say, another country to view your Web pages? Here are a few of the requirements.

1. You need to have Web pages to publish.

2. A Web server where the files can be placed must be available to you, and you need to learn the steps to put the files in place, either to create them in place or (more often) copy them into place after you develop and test them.

3. The permissions on the files need to be set so that any user anywhere can read them. Such file permissions are often referred to as *world readable*.

4. When someone requests your Web page, the server has to deliver it.

The details of exactly how these steps are performed vary from platform to platform. Under normal circumstances, you will only have to go through this entire setup process once. Thus, even though the procedure may be a bit technical, it is worth performing, as the rewards are great.

2.5.2 A Specific Example

In a UNIX environment, setting up a Web page usually involves creating a special directory in your home directory that contains all of your files to be published on the Web. This directory may contain subdirectories as well. Usually, the name of this special directory is fixed. Your systems administrator can tell you what the name of the directory should be on your system.

Suppose the name of the directory is `public-html`. You will need to use the `mkdir` UNIX command, which stands for "make directory," from within your home directory to create the `public-html` directory.

Since this directory must be accessible by others in order to permit them to read your Web pages, you will need to change the permissions on `public-html` to be world readable and world executable. You will also have to change permissions on your home directory so that it is world readable and world executable. This will allow others to access your `public-html` directory.

Be careful not to give anyone extra permissions on your home directory; do not make it world writable. Additionally, you should set permissions on private files sitting in your home directory so that only you can read them. Once the `public-html` directory is in place and the permissions on it and your home directory are set correctly, all your HTML files should be located there, or in subdirectories of the `public-html` directory.

When a Web page request is received for a URL that ends in a directory name, the Web server usually tries to load a default file. Again, the name of this default file can be determined by asking your systems administrator. Many installations use a file called `index.html`, which becomes your top-level Web page. Of course, you need to make this file world readable. If you use the default file name, users who know only your account name can access your Web page. How is this possible? Suppose you know that Sarah Conners' account name is `sarahc` and that her Web pages are served from a UNIX-based Web server called `pubpages.yikes.gov`. To access her pages, you could try the following URL:[5]

`http://pubpages.yikes.gov/~sarahc`

With a server default top-level page name of `index.html`, this is an abbreviation for the following URL:

`http://pubpages.yikes.gov/~sarahc/public-html/index.html`

In other words, the server automatically looks in the `public-html` directory and appends the default file name to a URL that does not contain a file name.

As you create new directories and Web pages, you will need to set the permissions on them so that others can access the directories and files. If you are creating a Web presentation with many HTML files (those with `.html` or `.htm` extensions) and graphics files (for example, those files ending in `.gif` or `.jpeg` extensions), you may be wise to organize your files into subdirectories of the `public-html` directory. For starters, you may want to create HTML, GIF, and JPEG subdirectories.

Summary: UNIX Web Page Setup

We will describe a typical scenario you might go through in setting up your Web page on a UNIX-based Web server. Suppose the directory where Web pages are placed is called `public-html` and the default file the server returns is called `index.html`. The following steps illustrate how to install this page. These steps assume you have read Appendix D or that you are familiar with

---

[5] The ~ and an account name combined are used to refer to the account owner's home directory. For example, ~sarahc refers to Sarah Conners' home directory.

the UNIX operating system. The commands must be executed in the order shown.[6] We preface each command with the UNIX prompt % and then give a brief explanation of what the command accomplishes.

1. `%cd`—Change to your home directory.

2. `%chmod og+rx ~`—Set the permissions on your home directory to be world readable and world executable.

3. `%mkdir public-html`—Create the directory `public-html`.

4. `%chmod og+rx public-html`—Set the permissions on the directory `public-html` to be world readable and world executable.

5. `%cd public-html`—Change directories to the `public-html` directory.

6. `%edit index.html`—Here, `edit` stands for your favorite text editor. The idea is to create a file called `index.html` and include the appropriate HTML code in it.

7. `%chmod og+r index.html`—Set the permissions on the file `index.html` to be world readable.

Once this sequence of steps, or a similar one depending on your local site, has been carried out, your `index.html` file should be ready to be viewed on the Web. Enter the URL for your page in the location area of the browser. For example, if your Web server is `www.chem.unlv.edu` and your account name is `shannon`, your URL would be something like

`www.chem.unlv.edu/~shannon/index.html`

If the page loads, congratulations, as you have just published your first Web page. If your page does not load, you may get an error indicating that the file protections are not set properly. Review the steps again and use the command `ls -l` to check that the protections on the necessary directories and the file `index.html` are set properly. Remember, as you install new files and subdirectories, you will need to set the permissions on these as you did on `index.html` and `public-html`, respectively.

EXERCISES 2.5 Web Page Installation

18. Create the necessary directories and files to install your Web page. For starting out, create a simple HTML document with a title of your name. Check to see that your page is accessible on the Web. Document any problems you have during the installation process.

---

[6] Technically, different sequences would work, but some steps depend on others.

19. What is the URL of your Web page? Does your server supply a default file name? If so, what is it? Can you use an abbreviated URL to access your Web page? If so, what is it? (When providing someone else with the URL of your personal page, it is often best to give them the shortest possible URL and thus minimize the chances for typing mistakes.)

20. Create new folders or directories to hold GIF and JPEG images. Set the protections on them so that images stored within them will be accessible.

2.6 Web Page Setup

Earlier you saw that each HTML document contains a head and a body. You will learn about the <HEAD> and <BODY> tags in detail here. In addition, we will examine colors, the <FONT> tag, the inclusion of hidden comments in an HTML document, and the methods for producing interesting backgrounds. Using the techniques described in the last section, you should be able to implement and test all of the HTML features discussed in this section. Keep in mind that many of our examples are HTML code fragments and not complete documents.

2.6.1 Head Tag

The head tag, <HEAD>, has no attributes. However, several tags can be included inside it. The most important of these is the title tag described earlier. A couple of others that you may find useful are described in the following paragraphs.

Basefont Tag

The basefont tag, <BASEFONT>, defines the font size to be used in the HTML document and may be included in the head of the document. It is also possible to use the basefont tag in other locations of an HTML document. Most browsers permit a range of font sizes. Seven different sizes are commonly available, with the sizes ranging from 1, which is the smallest, to 7, which is the largest. Figure 2.3 displays the seven different text sizes. Each browser renders an HTML document using a default font size, which is usually 3. To set the font size slightly larger for the overall document, you can use the SIZE attribute of the basefont tag, as follows:

```
<HEAD>
<BASEFONT SIZE = "4">
</HEAD>
```

You can also use a setting stated as

```
<BASEFONT SIZE = "+1">
```

FIGURE 2.3 The Seven Different Text Sizes That Browsers Typically Support.
A value of 1 indicates the smallest size and a value of 7, the largest size. The default is usually set to 3.

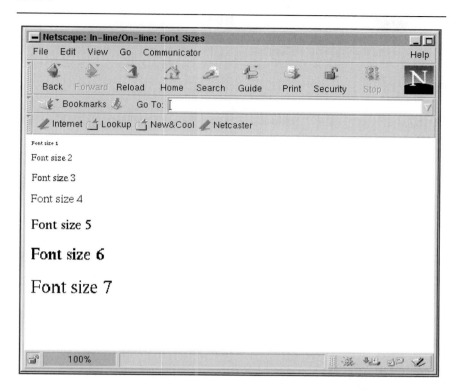

Notice the plus sign. This sets the font one size larger than the default. In typing this tag, it is common practice to omit the double quotes. The ending tag `</BASEFONT>` returns the font size to its default value. Note: when `<BASEFONT>` is used in the head of a document, the ending tag is usually omitted.

Base Tag

The base tag, `<BASE>`, is useful for setting some *global parameters* of an HTML document and may be included in the head of the document. A global parameter is an attribute that has an effect on the entire document. Before explaining how the base tag is used, you need to understand *absolute* versus *relative* URLs. An absolute URL is complete in that it contains all the components of a URL: the how, where, and what. For example, the following:

```
http://www.hospital.arizona.com/library/books/main.html
```

is an absolute URL. The how is

```
http
```

the where is

```
www.hospital.arizona.com
```

and the what is

```
library/books/main.html
```

By comparison, a relative URL, as the name implies, relates to some base URL and may be used in many different places in an HTML document. The default base URL of an HTML document is the URL of the document itself. However, the base URL can be changed using the `<BASE>` tag and its attribute `HREF`. For example, suppose we have the following code:

```
<HEAD>
<TITLE>Water Sports to Die For</TITLE>
<BASE HREF =
   "http://www.fishing.com/BOATS/outboard.html">
</HEAD>
```

in the HTML document[7] whose URL is

```
http://www.paloalto.gov/entertainment/water.html
```

Then all references to URLs in the file `water.html` would be relative to the base URL,

```
http://www.fishing.com/BOATS/outboard.html
```

instead of the default URL,

```
http://www.paloalto.gov/entertainment/water.html
```

More concretely, suppose a hyperlink in the file `water.html` referenced the URL

```
http://www.fishing.com/BOATS/inboard.html
```

---

[7] Notice that we have split the `HREF` over two lines. This is not significant. We have only done this so the reference did not extend into the margin of the book. In your files, you may include the expression on a single line. To minimize ambiguity, we normally break a line at a delimiter, such as a comma or an equals sign, or between attributes.

Having set the base URL to

```
http://www.fishing.com/BOATS/outboard.html
```

the referenced URL could be specified simply as `inboard.html`, since the prefix of the URL can be determined from the base URL. Why is this useful? If, for example, there are many references to URLs in the document collection found on the `www.fishing.com` server, their names can all be shortened.

Web addresses change frequently, and as explained here, the base tag can be used to simplify the updating of hyperlinks inside a file. If absolute URLs are hard-coded into HTML documents, then if a collection of documents moves to a new server, it may be necessary to edit all the files in the collection to update the URLs. This can result in a tremendous amount of editing. However, if relative URLs are used, it would probably only be necessary to update the base tag's `HREF` at the beginning of each document. We recommend using relative URLs where possible. An example should make this point clear.

Suppose a student at Winthrop University has the following absolute URL for her personal Web page:

```
http://www.winthrop.edu/JenniferJones/public-html/index.html
```

Figure 2.4 illustrates Jennifer's file structure, viewed as a tree. In her account, she has two top-level directories (or folders) called `private` and `public-html`. In her `public-html` directory, which contains her WWW material, she has four subdirectories and a file called `index.html`. The four subdirectories are `books`, `family`, `gif`, and `jpg`. The directories `gif` and `jpg` contain some of her graphics. The absolute URL for her Web page about her mother is

```
http://www.winthrop.edu/JenniferJones/public-html ↩
            /family/mom.html
```

(We use the symbol ↩ to denote that the line continues without any spaces.)

Suppose Jennifer's brother Jeff, who attends Tennessee Technical University, wants to include a picture of South Carolina in his HTML document called `sister.html`. Jeff could do this using an absolute URL of

```
http://www.winthrop.edu/JenniferJones/public-html/jpg/sc.jpg
```

He would have to type equally cumbersome URLs to include the picture of Rock Hill and also the picture of Jenny's favorite lake. In the interest of future portability, Jeff decides instead to use a base tag by including the following

FIGURE 2.4 **File Structure of Jennifer Jones' Computer Account.**

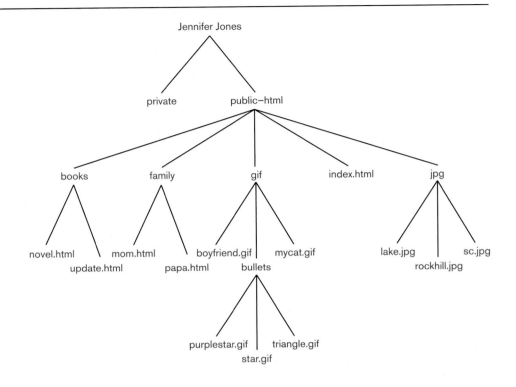

code in his file called `sister.html`.

```
<HEAD>
<TITLE>Jenny Jones: My Cool Sister</TITLE>
<BASE HREF = "http://www.winthrop.edu/JenniferJones ←
     /public-html/jpg/sc.jpg">
</HEAD>
```

The reference

```
http://www.winthrop.edu/JenniferJones/public-html/jpg/sc.jpg
```

can now be replaced by just `sc.jpg`. Similarly, a reference to

```
http://www.winthrop.edu/JenniferJones/public-html/jpg ←
     /lake.jpg
```

can be specified simply as `lake.jpg`. When Jenny graduates and moves her
document collection to a new Web server, Jeff only has to update the `HREF` in
his base tag, as opposed to changing all references to Jenny's files individually.

Jenny can also refer to her own document collection using relative URLs.
Suppose, for the sake of discussion, that `www.winthrop.edu` is a UNIX-based
Web server. Also assume that Jenny has not set the base tag in her `papa.html`
file. If Jenny wants to include a picture of Rock Hill in her `papa.html` file,
she could use an absolute URL of

```
http://www.winthrop.edu/JenniferJones/public-html ↩
      /jpg/rockhill.jpg
```

or she could use a relative URL of

```
                    ../jpg/rockhill.jpg
```

(Note: the symbol `..` is a way of moving up the directory structure one level.)
This URL is relative to the default URL for the file `papa.html`, which is

```
http://www.winthrop.edu/JenniferJones/public-html ↩
      /family/papa.html
```

In other words, the default URL for a file is just the URL of the file itself.

The expression `../jpg/rockhill.jpg` can be broken down as follows:

- "`..`" says go up one directory, which places us in the directory
  `/public-html`.
- `/jpg` says go into the directory `/jpg`.
- `rockhill.jpg` tells us the file name to include.

An expression such as

```
                ../gif/bullets/triangle.gif
```

in the file `update.html` could be used to refer to the triangle icon called
`triangle.gif`. To refer to the purple star bullet from within the file
`index.html`, Jenny could use the expression

```
            gif/bullets/purplestar.gif
```

We will have more to say about the base tag, and in particular its `TARGET`
attribute, when we discuss the concept of frames.

If possible, try to use all lowercase or all uppercase in your own URLs, since
this may help people avoid typing errors. Also, try not to use "underscore" (_)

and "dash" (-), and try never to use both of these in the same URL (or e-mail address). These symbols may be hard to distinguish on some monitors.

Meta Tag

The least well-understood and second most widely used (and occasionally abused) tag inside the head tag is the meta tag, <META>. This tag is used to include additional information about a document and can be used to pass additional information to a browser. There is no ending tag for <META>, and a document can have multiple <META> tags.

The attributes of the meta tag are NAME, CONTENT, and HTTP-EQUIV. You should include a modest list of keywords, say three to five, as the value of the CONTENT attribute. If someone is searching for a particular topic, your page may be returned if one or more of your keywords match their search request. Do not abuse the meta tag by expanding the number of items in the CONTENT attribute to ridiculous lengths. People who abuse the meta tag in this fashion are known as *spamdexers*, and there are many classic cases of this behavior.

For example, someone who has a Web page about woodworking might include the following:

```
<HEAD>
<META NAME = "keywords"
       CONTENT = "woodworking, cabinetmaking,
                   handmade furniture">
</HEAD>
```

2.6.2 HTML and Colors

Colors can help or hinder readers of your Web pages. There are two ways of defining colors in HTML documents. One involves straightforward color names, such as blue, cranberry, green, orange, red, and yellow. Many browsers have a list of predefined color names. Such lists are usually easy to find on-line. However, since different browsers have different lists and since the definitions of individual colors may vary from browser to browser, we recommend using the color numbering scheme. Although somewhat more complex, this scheme is better supported across different platforms.

A little computer numbering terminology is necessary first. A *bit* is either 0 or 1. The term bit stands for *bi*nary digi*t*. Bits are useful for counting in base two. Table 2.2 shows some binary numbers and their corresponding decimal values. In this text, we use the phrase "decimal number" to represent a base ten number from the set of natural numbers, 0, 1, 2, ... For example, 11101 in binary represents 29 in decimal, as follows:

$$(1 \times 2^4) + (1 \times 2^3) + (1 \times 2^2) + (0 \times 2^1) + (1 \times 2^0)$$

TABLE 2.2 Sample Decimal Numbers and Their Corresponding Binary Values.

Decimal	Binary
0	0
1	1
2	10
4	100
5	101
29	11101
255	11111111

equals

$$16 + 8 + 4 + 0 + 1 = 29$$

(Remember that any natural number raised to the zero power equals one.)

This explanation of binary numbers is really just a useful warmup. Colors in HTML documents are represented as *hexadecimal* numbers, which are numbers in base sixteen. Hexadecimal can also be considered a shorthand for representing four bits, since only four bits are needed to represent the numbers 0 through 15. Since there are only 10 base ten digits (0–9), we need six additional symbols in the hexadecimal number system: A, B, C, D, E, and F. Table 2.3 provides the values of these and some other hexadecimal numbers, with their corresponding decimal values. For example, 752 in hexadecimal represents the number 1874 in decimal, as follows:

$$(7 \times 16^2) + (5 \times 16^1) + (2 \times 16^0) = 1792 + 80 + 2 = 1874$$

You will be most concerned with hexadecimal numbers having two digits, because colors in HTML documents are represented by three two-digit hexadecimal numbers. Each of the two digits signifies the amount of one of three primary colors. In other words, a color is formed by mixing different amounts of red, green, and blue. The first two digits represent the red component, the next two the green portion, and the last two the amount of blue. This method of representing colors is called the *RGB color model*. We can view this as follows:

$$\underbrace{\text{digit1 digit2}}_{\text{red}} \underbrace{\text{digit3 digit4}}_{\text{green}} \underbrace{\text{digit5 digit6}}_{\text{blue}}$$

The first two digits, designated digit1 and digit2, represent the red component. For example, if digit1 = 0 and digit2 = 0, there is no red component. However, if digit1 = F and digit2 = F, the maximum possible red component is used. Since FF is the largest two-digit hexadecimal number (it represents 255

TABLE 2.3 Sample Decimal Numbers and Their Corresponding Binary and Hexadecimal Values, with Each Hexadecimal Number Written as Four Binary Digits.

Decimal	Binary	Hexadecimal
0	0000	0
1	0001	1
10	1010	A
11	1011	B
12	1100	C
13	1101	D
14	1110	E
15	1111	F
17	0001 0001	11
35	0010 0011	23
255	1111 1111	FF
1874	0111 0101 0010	752

in decimal), this is the maximum red we can specify using two hexadecimal digits. The green component is specified by digit3 and digit4, whereas the blue portion is given by digit5 and digit6.

As an example, 000000 means 00 or no red, 00 of green, and 00 of blue. This total absence of color is the color black. So, 000000 represents black.

It is common practice to preface these six-digit combinations by a # sign to denote that they represent a color. You can imagine the possible ambiguity arising from a six-letter color name consisting only of the letters A–F. Did the user want a color name, or a hexadecimal number to be interpreted as a color?

The color #FFFFFF represents bright red, bright green, and bright blue. This complete mix of these three colors yields white. Table 2.4 lists several color names and their corresponding hexadecimal values.

Desktop window systems allow the opportunity to download and install a *freeware* or *shareware* utility with a "color picker" to determine the hexadecimal value of any color you click on.

2.6.3   Body Tag

The body is the second and main part of every HTML document. The text and HTML code that goes between the body beginning and ending tags is rendered and displayed in the document area of the browser's window. The body tag, `<BODY>`, has a number of useful attributes that let you set some global parameters. The most interesting attributes deal with the document's text color and background color, and the properties of hyperlinks.

TABLE 2.4 **Some Colors and Their Corresponding Hexadecimal Representations.**

Color	Hexadecimal Value	Color	Hexadecimal Value
black	#000000	orange	#FFA500
blue	#0000FF	plum	#DDA0DD
chocolate	#D2691E	purple	#800080
crimson	#DC143C	red	#FF0000
gold	#FFD700	salmon	#FA8072
green	#00FF00	silver	#C0C0C0
gray	#808080	violet	#EE82EE
maroon	#800000	white	#FFFFFF
navy	#000080	yellow	#FFFF00

Text Color

The TEXT attribute is used to change the default text color for an entire document. (We will see how to override this setting in the next section on fonts.) Suppose you have a document and would like to use a maroon-colored text. The following HTML code shows how:

```
<BODY TEXT = "#800000">
```

A common mistake for beginners is the use of a text color that clashes with the background color. This makes your document hard to read and less desirable to visit. Make sure to select a background that goes well with your choice of text color. If possible, view the color combination on several different platforms.

Background Color and Tilings

Including effective colors (legible text and coordinated colors) in your Web pages can really improve appearance and navigability. Not only that, if you have a consistent color scheme running throughout your presentation, visitors will know they are still at your presentation as they select new hyperlinks. Conversely, if someone is surfing your Web pages and they click on a hyperlink taking them to a new page having a completely different color scheme, they will assume that they have left your pages. A carefully chosen color scheme can unite your pages and give them your own "look."

Two attributes to the body tag that let you add color to a Web page background are BGCOLOR and BACKGROUND. (The default is a gray background.) The BGCOLOR attribute is used to set the background of an HTML document to a single color. For example, the following HTML code sets the document background area to blue:

```
<BODY BGCOLOR = "#0000FF">
```

You can also use color names:

```
<BODY BGCOLOR = "blue">
```

As we explained earlier, we recommend you stick with hexadecimal numbers, since there are many shades of blue, and the one you see on your screen may not match the one another viewer might see on their system. Choosing a good background color is not easy and may require some experimentation. Just because your favorite color is aqua does not mean that it is the best choice for your Web pages. Select a color that is easy on the eyes and makes the text easy to read. White is usually safe. Black is difficult to use effectively. If you use a black background with white text, many users will have a problem printing your Web page. The black background color will not be printed, and the white text will not show up on white paper.

As for the BACKGROUND attribute, imagine holding up a postage-stamp-sized tile in front of your face. When you see an interesting tiled pattern on a Web page, it is usually created by taking postage-stamp-sized images and repeating them as necessary to fill in the document area in the browser's window. If you widen the document window, the pattern "expands" to fill it. Shrink the window and the pattern "contracts" to fill it.

The concept we are describing is called *tiling*. You may take any image and include it in your HTML document so that it tiles the background. The tiling is performed by the browser using a *tiling algorithm*. Abstractly, the tiler starts in one corner of the screen and horizontally lays tiles, which are copies of the image, until the right edge of the browser window is reached. If necessary, a tile is "cut." The tiler then moves down to the next horizontal row. The entire process is repeated, with the possibility that all tiles in the last row need to be cut.[8] Since computer scientists have figured out very efficient tiling algorithms, you do not notice any delay caused by the tiling procedure.

When using an image to tile a background, choose something that will not interfere with the legibility of your text. Common choices involve patterns of paper, clouds, and water. Marble textures are also popular. If you have an image called marble.jpg, the following HTML code would include it as a tiled background for you:

```
<BODY BACKGROUND = "marble.jpg">
```

As we said, be careful to choose a background pattern that makes your text easy to read. For example, a complex psychedelic image may be awesome,

---

[8] There are many presentations on the Web where good background graphics are available for free. This book's on-line Web presentation provides the URLs of several.

but it may not be suitable as a background. If no text color is clearly readable against that background, do not use it.

Some Web authors like to create *splash screen* effects by first loading in a color and then tiling a pattern over it. The following HTML code first loads in a color and then a background pattern:

```
<BODY BGCOLOR = "#008888" BACKGROUND = "dotblue.jpg">
```

Notice that when two attributes are used within the same tag, by convention a single blank space is used between the value of the first attribute and the name of the second attribute.

One reason for including the BGCOLOR attribute, even if you plan to use the BACKGROUND attribute, is that if someone has the automatic display of images turned off in their browser, the background image will not tile, but they will still get the background color. As a routine, the BGCOLOR attribute should be specified before the BACKGROUND attribute.[9]

Hyperlink Colors

Three attributes are used for changing the color of a hyperlink, where the color depends on the current *state of the hyperlink*. The three possible states are: unvisited, visited, and currently thinking of visiting. These are defined as follows:

LINK   Unvisited hyperlinks. The color value assigned to LINK sets the color for all unvisited hyperlinks in the HTML document.

VLINK   Visited hyperlinks. The color value assigned to VLINK sets the color for all visited hyperlinks, that is, hyperlinks the user has already explored.[10]

ALINK   A hyperlink the user is thinking of visiting. The "A" stands for *active hyperlink*. The color value assigned to ALINK sets the color of a hyperlink that the user has moused over and depressed the mouse button on. (This option is not supported by all browsers.)

Most browsers provide default colors for the three types of hyperlinks, and text-based browsers usually use underlining or reverse video to make hyperlinks more prominent on screen. However, many HTML documents specify all three attributes in the body tag. We encourage you to think carefully before you select colors that change the browser defaults people are used to seeing. Also, try to select colors that go well with your background and document text. Make sure your hyperlinks stand out; that is, do not set both the background color and LINK to red.

---

[9] In theory, the order should not matter, but it seems to on some systems.

[10] Your browser keeps a history file of where you have been, in addition to the cache and bookmarks, so it can determine which hyperlinks are to be treated as VLINKs.

Suppose you have an HTML document that has a white background. The following code specifies unvisited links as red, visited links as gray, and the active link as yellow:

```
<BODY BGCOLOR = "#FFFFFF"
      LINK = "#FF0000"
      VLINK = "#808080"
      ALINK = "#FFFF00">
```

Body Attributes Combined

Conceptually, it is straightforward to use all of the body attributes in combination. To do this effectively often requires some trial and error. When you see particularly effective colors used on someone's Web page, look at the document source and note the colors used. To create a white-colored document area with red text, along with green unvisited hyperlinks, orange visited hyperlinks, and purple active hyperlinks, use the following HTML code:

```
<BODY BGCOLOR = "#FFFFFF"
      TEXT = "#FF0000"
      LINK = "#00FF00"
      VLINK = "#FFA500"
      ALINK = "#800080">
```

2.6.4   HTML Font Colors

In the previous section, we saw how to set the text color of an entire document to any (single) color. In this section, we examine the font tag, `<FONT>`, that allows us to change the color of any portion of text. Modifying the color of a segment of text is easy using the `COLOR` attribute of the font tag. For example,

```
<FONT COLOR = "#0000FF">
I am going swimming today
</FONT>
```

changes the text "I am going swimming today" to blue. The preceding and succeeding text is unaffected. Only the text between the font beginning and ending tags is changed. When altering the color of small segments of text, check to make sure that the text is still readable when rendered. In general, do not use a large number of text color changes in a single HTML document.

2.6.5   Font Size

As previously stated, the basefont tag is commonly used in the head of a document to alter the font size for the entire document. The `SIZE` attribute of the font tag is typically used to change the font size of an individual part of a document.

One option, for example, makes the first letter of the first paragraph slightly larger than the rest of the text. This can be accomplished with the following code:

```
<FONT SIZE = "+3">W</FONT>elcome my friend.
```

Figure 2.5 illustrates the effect of this sample code. The remarks pertaining to the SIZE attribute for the basefont tag are also relevant here. For example, the SIZE attribute can have an absolute value of between 1 and 7.

2.6.6 Font Face

Most browsers also support a FACE attribute for the <FONT> tag, allowing you to specify a particular font type. For example,

```
<FONT FACE = "avantgarde">New font type</FONT>
```

FIGURE 2.5 Illustration of Enlarging the First Character in the Beginning Paragraph of an HTML Document.

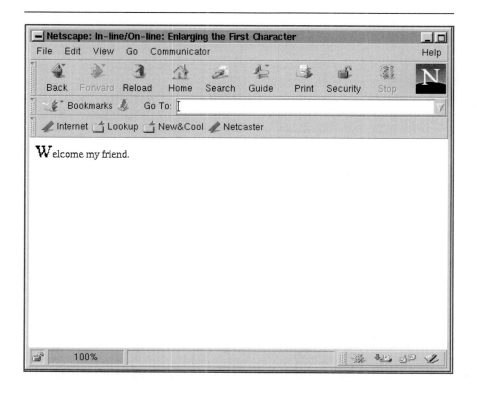

specifies that the phrase "New font type" should be displayed in the Avantgarde font. However, not all browsers support the same font families, and various fonts are rendered in different sizes. We therefore recommend using only a limited number of different font types. Otherwise, the spacing and font type you see may be completely different than what someone else sees when viewing your Web page.

2.6.7 HTML Comments

The comment feature provides you with a way to document your HTML files and make notes to yourself. Those who have written program code realize the importance of documenting the process concurrently with program construction. The same can be said for Web page design. Without such notes, it is unlikely you will be able to remember all the important details that went into a Web page's design.

The comment tag is not `<COMMENT>`; it is the set of symbols   `<!--`   for the beginning tag and   `-->`   for the ending tag. A comment, properly included, does not change the appearance of your Web page.

To use the comment tag, place the comment text between the pairs of dashes in the tag. For example,

```
<!-- This is a comment. -->
```

The browser will not interpret (i.e., display) the text between the pairs of dashes.

Do not include any embedded HTML code in commented text, since the results are unpredictable. Making a solid line of dashes may also cause unpredictable results in some browsers. Use periods or asterisks instead to get a similar effect. In the exercises, we ask you to experiment with how your browser handles HTML embedded within a comment declaration.

Here is an example of how the comment tag might be used.

```
<!-- How to paint a house in seven easy steps. -->
<!-- Written in July of 1997.  Gretchen von Gelder -->
<!-- Most of the description is from my file Home-notes. -->


Hi.  Welcome to the house painting scrap book.  A gallon
of paint covers about 400 square feet.  So, if you live
in a box that is 10 yards on a side, you would need
<!-- (4 x (10 x 3) ^ 2)/400 = 9 -->
9 gallons of paint to cover it.
```

(We have formatted the text inside the comment so that it looks good on the printed book page.) Including a comment like the last one helps you recall the actual calculation that you made. If some user sends you e-mail in January of 1999 asking how you figured on 9 gallons of paint, you can simply consult your comment to review the math. The notes at the top of the document remind you what the HTML code is about, who wrote it, and when. It is generally a good idea to include this information in an HTML file, or any on-line file for that matter. The output resulting from this HTML code is shown in Figure 2.6.

You could also embed a copyright notice at the top of a document. For example,

```
<!-- copyright (c) 1997, Mark Jackson -->
```

With the basic principles and HTML tags described thus far, you should be able to create some interesting Web pages. Remember to comment them so you can easily recall what you did, why, and when.

EXERCISES 2.6 Web Page Setup

21.   Provide the head specification for an HTML document in which the entire document's font size is set to a value two less than the default size.

**FIGURE 2.6 Display of a Web Page That Includes Comments.
Note that the comments are not rendered by the browser.**

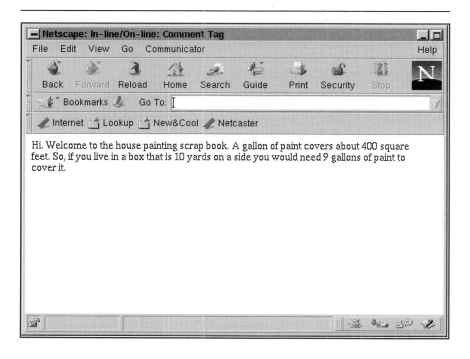

22. Provide an absolute URL that Jennifer Jones could use to include her South Carolina image in her `novel.html` file. What is a relative URL she could use to do the same thing?

23. Document two widely published cases involving spamdexers. (Hint: Use a search engine such as Yahoo!)

24. What are the decimal equivalents of the following hexadecimal numbers: 16, 22, AD, CB, FG, 79, and FF?

25. What are the probable hexadecimal values for the following colors: black, cranberry, lime, orange, paleblue, royalblue, and white?

26. Andrew "Thumbs" Michaels is typing in a color specification, intending to use a color name not a hexadecimal number. He "accidentally" hits the shift-key while bumping the 3-key, so the code starts out with a # sign. He then keys in his desired color name, making at most four typos. Are there any color names he could have botched that would result in a valid hexadecimal number? As an example, if he meant to key in "beige" but made three typos $i \mapsto a$, $g \mapsto a$, and accidentally appended an a, the result in capitals would be BEAAEA, a valid hexadecimal number.

27. Report the URLs for three different Web pages that have cross references providing hexadecimal values for color names.

28. Can you locate a Web page that allows you to enter a color name and then have it return the corresponding hexadecimal representation? How about vice versa?

29. Are color names case sensitive? How about letters in hexadecimal color representations? (Notice that we have always used lowercase letters for color names and capital letters in hexadecimal numbers.)

30. Write a body tag specification to produce a salmon-colored background. Do this once using the color name "salmon" and once using the hexadecimal representation for salmon.

31. Suppose you did not have (and could not find) a table to look up hexadecimal values for a given color. Describe how you could obtain an approximate value just using HTML code and your browser.

32. What colors do the following hexadecimal patterns probably represent: #00FF00, #FF0000, #FFFF00, #FA5723, and #00BC51?

33. Write a `<BODY>` tag specification to generate a background that is tiled using the pattern

```
http://www.herewego.com/backgrounds/droplets.gif
```

If you use an absolute URL, then every time your page loads, this background pattern would need to be loaded from another server. This is considered bad practice and is certainly time consuming.[11] Suppose instead that you copy the image to your disk area. Now write a specification using a relative URL to generate the same background.

34. Write HTML code for a body tag in which you have gold unvisited hyperlinks, silver visited hyperlinks, and bronze active hyperlinks. Do you think this is a good combination to use? Why or why not?

35. Design a single-screen Web page about your favorite animal, using the tags described so far. That is, only use tags presented thus far in the book and do not omit any tags. Use a sensible set of colors. For example, if you love turtles, use a green background with a variation on yellow as a text color, or whatever else makes sense to you. Carefully think about your design and style when you code the page. Make sure to comment the code, explaining why you chose the colors you did.

36. Design a single-screen Web page about your favorite beverage, using the tags described so far. (See previous exercise for further details.)

37. Describe three scenarios indicating when it would be a good idea to include a comment in an HTML file.

38. How does your browser handle embedded HTML tags within a comment? Does it try to render them?

2.7   HTML Formatting and Hyperlink Creation

At this point you should be creating and viewing simple HTML documents. This section introduces you to four more HTML tags. They are `<P>` for paragraph, `<Hi>` for heading i, `<A>` for anchor, and `<IMG>` for image. By using these tags, you can make your pages more readable, interesting, and polished. The most interesting of these tags is the anchor tag; it allows you to create hyperlinks to other Web pages.

HTML tags describe the desired structure of a Web page, rather than exactly how it should look. For example, HTML tags identify emphasized text, headings, lists, and so on, not items like 11-point font and 0.75-inch margin. Because various browsers will render the same HTML code differently, it is important not to try to force a very specific layout on a Web page. Doing so may cause the page to look fabulous using one browser and awful using another. Keep this point in mind as you increase your repertoire of HTML tags.

---

[11] We know the Jennifer Jones example contained some code like this. For illustration purposes of URLs and since we did not expect to get caught by readers too often, we decided to leave it.

2.7.1 Paragraph Tag

The paragraph tag is used to break the text into paragraphs. Most browsers place an empty vertical space between paragraphs so they stand apart from each other. To designate a block of text as a paragraph, enclose it within the paragraph beginning and ending tags: <P> and </P>. The ending tag is considered optional since (most) browsers assume that the current paragraph ends when the browser encounters the next <P>. Nevertheless, we recommend treating <P> as a paired tag, even though it is not mandatory. The following HTML code illustrates the use of the paragraph tag:

```
<P>
This is the title sentence for the first paragraph.
You will notice that after learning a lot of HTML tags,
they are a little hard to keep straight.
This concludes paragraph one.
</P>

<P>
This is the start of paragraph two.
One good way to learn HTML tags is to practice using
them by creating Web pages.  In this way, they are
easier to remember.
This concludes paragraph two.
</P>

<P>
</P>

Yoga is an ancient art form.
```

The last paragraph tag leaves a blank vertical space between the sample text and the succeeding HTML elements. One rendering of the HTML code for the paragraph example is depicted in Figure 2.7.

Some Web authors will put a series of paragraph tags adjacent to one another to skip some vertical space. For example,

```
<P> <P> <P> <P>
```

will cause most browsers to leave approximately 1 inch of vertical space. This is the type of HTML programming we generally recommend against. Do not try to force a precise spacing, because it may look exactly as intended on your

FIGURE 2.7 Sample Rendering of the Paragraph Tag Example.

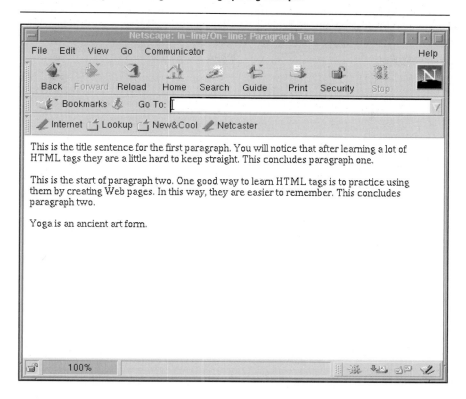

desktop, but it could look quite different to someone else using a different window size and/or browser.

2.7.2 Heading Tags

In many forms of writing, it is common to include section headings to provide the reader with a sense of the document's structure. When viewing any page of text, not just material on the Web, the first things most readers notice are headings and subheadings. As an example, glance at any page of this book containing a heading and observe that it stands out from the surrounding text.

Most browsers support a hierarchy of six levels of HTML headings. The beginning tag for heading i is <Hi>, where i can be any value from 1 to 6. The corresponding ending tag, as expected, is </Hi>. The largest heading is <H1> and the smallest is <H6>. Note that this is the reverse of the way sizes are specified in the font and basefont tags.

The heading tags are very useful for dividing a document into sections: the more important the section, the larger the heading tag. Subsections are usually less important and so receive smaller headings. Here is sample HTML code that illustrates the use of the heading tag.

```
<H1>Complete Sentences</H1>

Most of us would agree that well-written English ...

<H2>Fragments and Run-on Sentences</H2>

Keep in mind that rambling endlessly can lead to ...

<H3>The Phrase Fragment</H3>

Phrase fragment description goes here.

<H3>The Appositive Fragment</H3>

Appositive fragment description goes here.

<H4>Examples</H4>

There are two types of examples: sentence fragment and
sentence complete.

<H5>Sentence Fragment</H5>

<H5>Sentence Complete</H5>

<H6>Copyright, English Grammar for the Rest of Us</H6>
```

Figure 2.8 illustrates how one browser renders the example heading code. The browser determines the exact style of the headings; that is, the font size for each heading, whether the heading is boldface, whether the heading gets numbered, and so on. Notice that as the topics become more specialized, the heading sizes become smaller. Three or fewer levels of headings will suffice for most writing.

It is a fairly common practice to use level 5 or 6 headings for copyright notices and disclaimers, as Figure 2.8 shows. These headings are usually too small for anything else.

FIGURE 2.8 Sample Rendering of the Heading Tag Example.

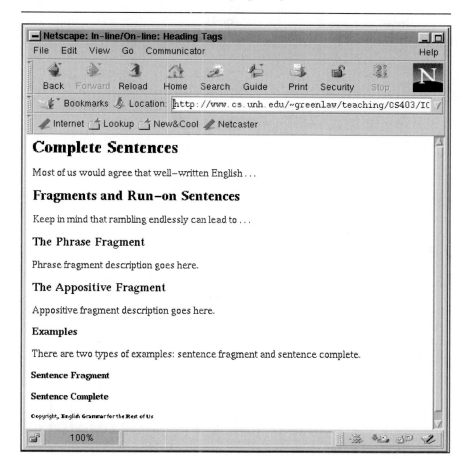

One important attribute of the heading tag is ALIGN, which can have values of left, center, or right. For headings that are not as long as the width of the document area, the ALIGN attribute has the expected effect. Caution should be exercised here. If the document is viewed in a small window, the ALIGN attribute may produce an undesirable appearance.

2.7.3 Anchor Tag

The anchor tag, <A> and </A>, is the mechanism by which hyperlinks are placed in hypertext documents. Its syntax is more complicated than that of most other tags. The term "anchor" is used because it indicates the static positioning of a hyperlink. In this section, we explain how to create clickable text hyperlinks, clickable images, `mailto` hyperlinks, and hyperlinks that point inside a document. We also provide some guidelines on hyperlink construction.

The three basic parts of a hyperlink are:

- The beginning and ending tag pair `<A> ··· </A>`.

- The `HREF` attribute that specifies the URL of the page to be loaded when the hyperlink is selected.

- The text (or graphic) that appears on-screen as the active link.

Clickable Text Hyperlinks

Consider the following example:

```
<A HREF = "http://www.usa.gov/documents/ ↩
          whitehouse.html">White House</A>
```

We have used the `HREF` attribute of the anchor tag to create a hyperlink labeled "White House." Several items about this hyperlink are worth pointing out. First, most graphical browsers will change the text color and underline the "name" of the hyperlink, which in this case is White House. When someone clicks on this phrase, the file `/documents/whitehouse.html` is requested from the server `www.usa.gov`. Second, notice that we did not leave any blank spaces before or after the text "White House." If you leave blank spaces, the hyperlink underline is drawn too wide. For example, the hyperlink,

```
<A HREF = "http://www.usa.gov/documents/ ↩
          whitehouse.html">   White House   </A>
```

results in the following underlining:

<u><code>White House</code></u>

Third, the phrase White House is short and descriptive. Most users would be able to guess quickly what type of information they would be retrieving by selecting this hyperlink. In your Web pages, use short, informative hyperlink names. Finally, notice that in our example, we used an absolute URL. It is possible to use relative URLs, as well.

Clickable Image Hyperlinks

The principles behind creating a clickable image are the same as for creating a clickable text hyperlink. The type of hyperlink we describe here consists of a single image, for which one mouse click returns an HTML document. (It is possible to define images that can have a number of mouse-sensitive areas, each with the potential to return a different Web page.) The idea is simply to replace the clickable text with an image. As an example, consider using the file `wheelbarrow.gif`, which contains a 50-by-50 *pixel* image called "Under Construction."

```
<A  HREF = "http://www.usa.gov/wogulis/notready.html">
    <IMG  SRC =  "wheelbarrow.gif"
          ALT = "Under Construction"
          HEIGHT = "50"
          WIDTH = "50">
</A>
```

If a user clicks on the wheelbarrow image, the document `notready.html` will be loaded. Browsers will typically draw a highlighted border two pixels wide around the image. It is not always completely obvious that an image is a hyperlink, so it is sometimes worthwhile to add text to alert the reader to this fact. This requires some thought in order not to defeat the purpose of using the image.

Mailto Hyperlinks

It is common practice to add a `mailto` hyperlink to a Web page. This provides a convenient method for someone viewing your page to send you e-mail. Suppose Pascal Leno has an e-mail address of

```
leno@oli.rustica.it
```

and wants to include a `mailto` hyperlink labeled "Contact Pascal" on his Web page. The following code does it:

```
<A  HREF = "mailto:leno@oli.rustica.it">Contact Pascal</A>
```

Note that the syntax for the `mailto:` URL is different from the `http://` URL because the double slashes are not allowed.

When the user clicks on the hyperlink "Contact Pascal," a mail dialog box (this is not your usual mail client bundled with the browser) will be launched, with the *To* field filled in with `leno@oli.rustica.it`. All the user has to do is complete the remainder of the message and send it. This presupposes that you filled in your name and e-mail address under options or preferences for your browser, or in a public room situation, that you have verified the settings and updated them to point to you.

Intradocument Linking

Another important attribute of the anchor tag is NAME. The NAME attribute lets you create a hyperlink to any part of your document, rather than just the beginning. That is, any portion of the document can automatically be displayed at the top of the browser's document area. This is particularly useful if you have a long Web page and would like users to be able to jump to various sections of it without scrolling. Many Web authors provide an index at the top of a page, with hyperlinks that jump to other parts of the document. Using NAME hyperlinks means you do not have to break the document down into pieces to allow the reader easy and rapid navigation.

How are NAME hyperlinks created? The following HTML code demonstrates the process:

```
Welcome to the Lemonade Parade.  We serve the best
lemonade anywhere.  Each flavor has a history of its
own.  We offer
<A HREF = "#blueberry">Beautiful Blueberry</A>,
<A HREF = "#cherry">Cherry Delight</A>, and
<A HREF = "#lemon">Luscious Lemon</A>.
...
<A NAME = "blueberry">
<H3>Beautiful Blueberry</H3>
</A>
...
<A NAME = "cherry">
<H3>Cherry Delight</H3>
</A>
...
<A NAME = "lemon">
<H3>Luscious Lemon</H3>
</A>
...
```

A rendering of this code is shown in Figure 2.9. Note that if the hyperlink is in the same file, as in our case, the URL does not need to be specified. You can begin the HREF value with the # symbol, as shown here.

FIGURE 2.9 Sample Rendering of the Code Used to Illustrate the NAME Attribute of the Anchor Tag.

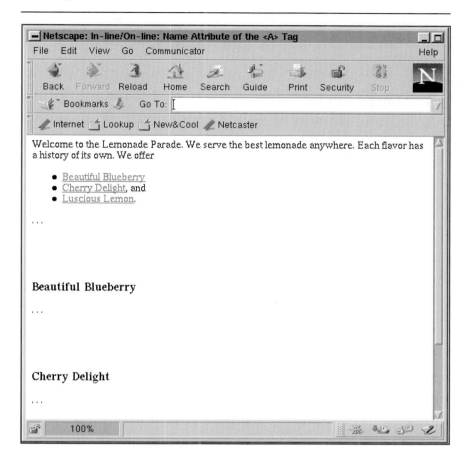

The NAME attribute is used to label three separate sections of the document. Notice that between the beginning <A> and ending </A>, we have included the section heading (in this case, for the particular flavor of beverage described). You should always include at least one line of text between the anchor tags.

You can create a hyperlink to a label by using the URL for the file in which it is contained, followed by the # symbol and then the label name, as follows:

```
HREF = "http://systemname/docpath#labelname"
```

where `systemname/docpath` is the URL, and `labelname` is the actual label used.

What is the effect of selecting one of these hyperlinks? If a user clicks on the Luscious Lemon hyperlink, for example, the cursor immediately moves down to the section labeled Luscious Lemon. Here, this section is contained in the next screen. This feature obviates the need for scrolling and provides the user with a more convenient method of accessing different parts of a long document.

Hyperlink Creation Guidelines

Do:

- Carefully choose the text that goes into a hyperlink.
- Create hyperlinks that are interesting.
- Keep the hyperlink text short and descriptive.
- Create hyperlinks that read well together, even without any intervening text.

Don't:

- Use hyperlinks that may split over two or more lines.
- Use additional underlining near hyperlinks.
- Put two or more hyperlinks side by side.
- Use full-line-length hyperlinks.
- Use such phrases as "click me," "click here," or "click now."

Figure 2.10 illustrates the Don't list.

2.7.4 Image Tag

The image tag, $\langle$IMG $\rangle$, is used for including in-line images in HTML documents. An example of the use of the tag is:

```
<IMG  SRC = "wheelbarrow.gif"
      ALT = "Under Construction"
      HEIGHT = "50"
      WIDTH = "50" >
```

We will explain the meaning of this code, a number of attributes of the image tag, and several style issues.

FIGURE 2.10 Examples Illustrating Poor Choices of Hyperlinks.

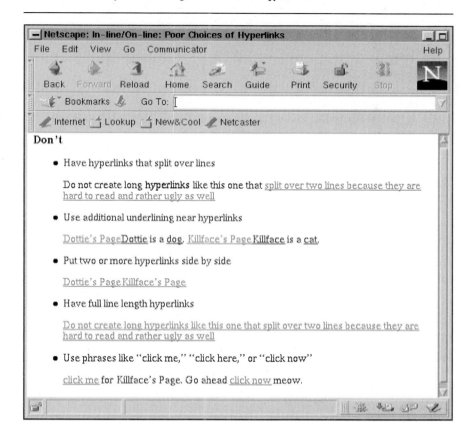

The most important attribute of the image tag is SRC, which is used to specify the image to be displayed. Any type of image can be specified, using either a relative or an absolute URL, where the relative URL would relate to the document in which the image appears. The most common image types on the Web are gif, jpg, and png. To include the image friend.gif in a file located in the same directory, you could use the following HTML code:

```
<IMG SRC = "friend.gif">
```

This is the minimum amount of code you can use to include an image.

When a browser retrieves a Web page, it does not automatically get the images that go along with that page. Each image must be retrieved separately. To render the document on-screen, the browser must know the sizes of the images. The browser can obtain these sizes either from code entered by the image's developer (as we recommend), or by reading the sizes as the images are brought over. The latter case takes longer, because the browser has to do the interpreting. So, a Web page will render more quickly if you include the size, using the HEIGHT and WIDTH attributes of the image tag, for each image.

Image dimensions are expressed in *pixels*. Suppose the picture friend.gif is 60 pixels wide and 90 pixels high. The following code would include the image in the Web page and would specify its dimensions to the browser:

```
<IMG SRC = "friend.gif"
     HEIGHT = "90"
     WIDTH = "60">
```

While the order in which you specify the HEIGHT and WIDTH attributes is not significant, we usually specify them in the order shown. On the other hand, a browser will list image dimensions as $x \times y$, where x is the WIDTH of the image and y is the HEIGHT.

When a browser parses this HTML code, it can determine how much space to leave for the image. Thus, the surrounding text can be rendered immediately. The browser does not have to wait until the image itself arrives. This is why you often see all of the text in a page long before all of the images are rendered.

How can you determine the size of an image? Most browsers have a "document info" menu item where you can find the dimensions of an image (in pixels). Of course, if you create an image yourself, you can record its size at that time.

The HEIGHT and WIDTH attributes can also have percentages as values, allowing them to be used to scale an image relative to the size of the browser's window. For example, the following code produces a version of friend.gif that occupies 50 percent of the browser's window in each dimension.

```
<IMG SRC = "friend.gif"
     HEIGHT = "50%"
     WIDTH = "50%">
```

You can create some interesting scaling effects by using percentages. However, when you scale an image downward, you do not reduce the amount of disk space required to store the image. Thus, it is not possible to create a reduced-size *thumbnail sketch* by using percentage values for the HEIGHT and WIDTH attributes.

The image tag has another interesting attribute known as ALT, which is short for "alternative." The value of ALT is a text string that usually describes the image in words. In our wheelbarrow.gif example, we used the words "Under Construction" as a value of the ALT attribute, because wheelbarrow.gif contains the picture of a wheelbarrow at a construction site. You have probably seen many Web pages that have "Under Construction" images. If a browser has images turned off, or is text-only, the words in the ALT attribute will be displayed on-screen where the image would have been. Obviously, the size of the image will not exactly match the text replacing it. Most Web authors do not worry about this detail, as most of their effort goes into making the pages look good with the images.

In the latest versions of some browsers, when the user mouses over an image, the text in an ALT attribute is displayed in the form of a *tooltip*. This is usually a light-colored dialog box that is just large enough for the text. For pages that contain a lot of images, the tooltips can become a distraction. This has prompted some users to stop including ALTs. Remember, however, that the purpose of ALTs is to serve those users who are unable to display images, or who are using text-based browsers. In such cases, the text of an ALT can provide the reader with some continuity.

Occasionally, square brackets have been used around the ALT attribute value. This mimics the convention adopted by some text-based browsers. However, on some browsers on some systems, this can create a problem, so we recommend against this practice. For example, do not use:

```
ALT = "[Under Construction]"
```

Instead use:

```
ALT = "Under Construction"
```

As another attempt at presenting an image in a text-based browser, some extremely clever Web authors place an ASCII graphic of the original image in the ALT tag. This is interesting, but very time consuming unless you can locate a free copy of the ASCII graphic you want to include.

With the material we have covered so far, you should be able to include images on your Web pages. Remember to set protections on your image files, similar to what you did for your index.html file.

EXERCISES 2.7 HTML Formatting and Hyperlink Creation

39. Create an HTML document that contains the paragraph tag. What happens if you include five paragraph tags in a row, with no intermediate text? Does the paragraph tag have any attributes? If so, explain and provide examples.

40. Create an HTML document that uses heading tags. As an example, the document could outline a recent or planned trip. Focus on the outline and on the use of the heading tags, rather than on the details of the trip.

41. Does your browser support six different-sized headings, or are some of them the same size? What happens if you try to close an <H1> tag with </H2> instead of </H1>? Report the URL of a Web page that makes effective use of headings.

42. Explain what happens if you use an ALIGN value of center in an <H4> tag that surrounds text that is wider than the browser's document area.

43. Create an HTML document containing hyperlinks to three Web pages about Italy.

44. Create a Web page containing a hyperlink to itself. What happens when you click on the hyperlink? Explain.

45. Produce a Web page that has hyperlinks to three of your friends' Web pages.

46. Design a Web page that contains a mailto hyperlink to yourself. Test it. Were you able to send yourself e-mail?

47. Create a Web page that can be used to link in the assignments for this course.

48. Code an HTML document that contains a clickable image.

49. Locate an "Under Construction" image and include it on a Web page. In addition to SRC, be sure to use the ALT, HEIGHT, and WIDTH attributes of the image tag.

50. Experiment with scaling an image. How small can you scale an image before it becomes "fuzzy"? Does this depend on the image quality with which you started? How large can you expand an image? What happens to the quality of the image as it gets larger?

The Internet

3

3.1 Introduction

Even though you may have been using the Internet for a while, you may not know very much about how it works, its culture and history, and some of its uses. The goals of this chapter are therefore to:

OBJECTIVES

- Present a "definition" of the Internet and some interesting facts about it.

- Discuss the history of the Internet (in the form of a timeline classifying some of the important events in the development of the Internet).

- Provide an intuitive idea of how the Internet functions.

- Discuss Internet congestion and what, if anything, can be done to address it.

- Teach you about the Internet's rich culture.

- Talk about the importance of business in the Internet setting.

- Define some of the issues pertaining to *collaborative computing* and its relationship to the Internet.

3.2 The Internet Defined

Many people, including the president and vice president of the United States, refer to the Internet as the *Information Superhighway*. The extended metaphor with cars and freeways is a useful one.

The following definition of the Internet was formulated by the *Federal Networking Council (FNC)*, which passed a resolution on October 24, 1995, defining the term. The resolution states:

> The Federal Networking Council (FNC) agrees that the following language reflects our definition of the term "Internet."
>
> "Internet" refers to the global information system that—
>
> 1. is logically linked together by a globally unique address space based on the Internet Protocol (IP) or its subsequent extensions/follow-ons;
>
> 2. is able to support communications using the Transmission Control Protocol/Internet Protocol (TCP/IP) suite or its subsequent extensions/follow-ons, and/or other IP-compatible protocols; and
>
> 3. provides, uses or makes accessible, either publicly or privately, high level services layered on the communications and related infrastructure described herein.

This definition of the Internet can be simplified to:

The Internet is a global system of networked computers together with their users and data.

To explain, the system is global in the sense that people from all over the world can connect to it. Also, since the users of the Internet have developed their own culture, they are a defining factor of the Internet. Finally, without the possibility of accessing data or personal information, no one would be excited about connecting to the Internet.

The concept of being able to access information quickly and easily and to communicate more easily and quickly led to the vision of the Internet. Thirty years ago, information exchange and communication took place via the "backroads"—regular postal mail, a telephone call, a personal meeting, and so on. Today, they take place nearly instantaneously over the Internet. The history section of this chapter describes the evolution of the Internet into today's Information Superhighway.

3.2.1 The Information Superhighway

Expanding on the freeway metaphor, with cars, there are various levels of knowledge; learning to drive is easy, and it is all you really need to know about cars. This is like learning to surf the Internet. In the course of driving, you learn about highways, shortcuts, and so on, and using the Web is very similar; that is, with practice, you will learn where and how to find things.

Also, in driving, you can go another step and learn how an engine works and how to do routine maintenance and repairs, such as oil changes and tune-ups. On the Web, the equivalent is to learn how Web pages are put together, which you have already started to do.

A still deeper level of involvement with cars is learning how to do complex repairs, or to design and build them. Not many people pursue cars to this stage. On the Web, a similar level of involvement is writing software, either building *applets* in a language such as Java, or developing more general-purpose tools for others to use in navigating the Web. Again, only a limited number of people aspire to this level.

Today, the Information Superhighway is in place, but for many people, the mysteries surrounding it involve where to go and how to travel. Like traveling a highway in a foreign country and being unable to read the road signs, navigating the Information Superhighway can be frustrating and time-consuming without the right knowledge and tools.

Consider that there are many ways to travel sidewalks, roads, and freeways to get to where we want to go. We can take a bicycle, a bus, a car, or a pair of in-line skates. Similarly, there are many ways to use the Internet to send and retrieve information. These include, but are not limited to: e-mail, *file transfer*, *remote login*, and the Web. New methods of using the Internet will probably be conceived and developed in the near future, and existing methods will be improved.

3.2.2 Interesting Internet Facts

Another way to gain insight about the Internet is to examine a few statistics about it.

- Each day, approximately 33,000 new users go on-line.

- As of August 1996, there were 72,600,000 users worldwide. Extrapolating growth to June 1, 1998, yields an estimate of 90,000,000 users.

- About one-third of Internet users are female.

- In January 1997, the number of Internet hosts exceeded 16,000,000.

- Internet connections were available in 177 countries, as of March 1997.

- Every three months, the number of World Wide Web sites approximately doubles.

EXERCISES 3.2 The Internet Defined

1. Update the statistics presented in this section to be as current as possible. Provide URLs for your sources.

2. It has been said that the number of Internet users doubles every year. How long can this process go on, assuming the earth's population is six billion? Explain why the growth will slow down at some point.

3. What conclusions can you draw from the Internet growth statistics?

4. Present four more interesting facts about the Internet. Explain why each is interesting to you. Provide references for your sources.

3.3   Internet History

The history of the Internet is best explained via a timeline, as shown in Figure 3.1. We have included events that were important and required innovation, as well as other interesting and related items. For each item mentioned on the timeline, we provide a brief synopsis. While the timeline begins in 1969, we present some general comments on the 1960s, for background. The history of the Internet is fascinating both for itself and as a case study of technological innovation.

FIGURE 3.1 Timeline Illustrating Important Dates in Internet History.

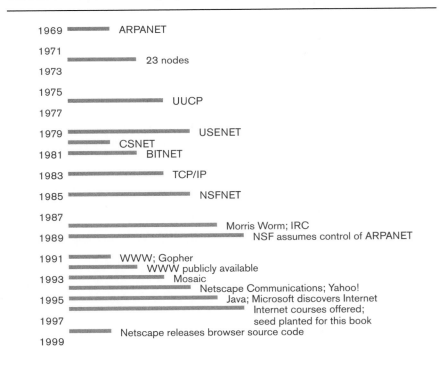

3.3.1 1960s Telecommunications

Essential to the early Internet concept was *packet switching*, in which data to be transmitted is divided into small packets of information and labeled to identify the sender and recipient. The packets were sent over a network and then reassembled at their destination. If any packet did not arrive or was not intact, the original sender was requested to resend the packet. Prior to packet switching, the less efficient *circuit switching* method of data transmission was used. In the early 1960s, several papers on packet switching theory were written, laying the groundwork for computer networking as it exists today.

ARPANET, 1969

In 1969, Bolt, Beranek, and Newman, Inc., (*BBN*) designed a network called the *Advanced Research Projects Agency Network* (*ARPANET*) for the United States Department of Defense. The military created ARPA[1] to enable researchers to share "super-computing" power. It was rumored that the military developed the ARPANET in response to the threat of a nuclear attack destroying the country's communication system.

Initially, only four nodes (or *hosts*) comprised the ARPANET. They were located at the University of California at Los Angeles, the University of California at Santa Barbara, the University of Utah, and the Stanford Research Institute. The ARPANET later became known as the Internet.

3.3.2 1970s Telecommunications

In this decade, the ARPANET was used primarily by the military, some of the larger companies, such as IBM, and universities (for e-mail). The general population was not yet connected to the system and very few people were on-line at work.

The use of *local area networks* (*LANs*) became more prevalent during the 1970s. Also, the idea of an *open architecture* was promoted; that is, networks making up the ARPANET could have any design. In later years, this concept had a tremendous impact on the growth of the ARPANET.

Twenty-Three Nodes, 1972

By 1972, the ARPANET was international, with nodes in Europe at the University College in London, England, and the Royal Radar Establishment in Norway. The number of nodes on the network was up to 23, and the trend would be for that number to double every year from then on. Ray Tomlinson, who worked at BBN, invented e-mail.

---

[1] Over the next several decades, ARPA flip-flopped between being called ARPA and *DARPA* (*Defense Advanced Research Projects Agency*). As of 1998, the name is DARPA.

UUCP, 1976

AT&T Bell Labs developed *UNIX to UNIX CoPy* (*UUCP*). In 1977, UUCP was distributed with UNIX.

USENET, 1979

User Network (*USENET*) was started by using UUCP to connect Duke University and the University of North Carolina at Chapel Hill. Newsgroups emerged from this early development.

3.3.3 1980s Telecommunications

In this decade, *Transmission Control Protocol/Internet Protocol* (*TCP/IP*), a set of rules governing how networks making up the ARPANET communicate, was established. For the first time, the term "Internet" was being used to describe the ARPANET. Security became a concern, as viruses appeared and electronic break-ins occurred.

The 1980s saw the Internet grow beyond being predominantly research oriented to including business applications and supporting a wide range of users. As the Internet became larger, the *Domain Name System* (*DNS*) was developed, to allow the network to expand more easily by assigning names to host computers in a distributed fashion.

CSNET, 1980

The *Computer Science Network* (*CSNET*) connected all university computer science departments in the United States. Computer science departments were relatively new, and only a limited number existed in 1980. CSNET joined the ARPANET in 1981.

BITNET, 1981

The *Because It's Time Network* (*BITNET*) formed at the City University of New York and connected to Yale University. Many *mailing lists* originated with BITNET.

TCP/IP, 1983

The United States Defense Communications Agency required that TCP/IP be used for all ARPANET hosts. Since TCP/IP was distributed at no charge, the Internet became what is called an *open system*. This allowed the Internet to grow quickly, as all connected computers were now "speaking the same language." Central administration was no longer necessary to run the network.

NSFNET, 1985

The *National Science Foundation Network* (*NSFNET*) was formed to connect the *National Science Foundation's* (*NSF's*) five super-computing centers. This allowed researchers to access the most powerful computers in the world, at a time when large, powerful, and expensive computers were a rarity and generally inaccessible.

The Internet Worm and IRC, 1988

The virus called *Internet Worm* (created by Robert Morris while he was a computer science graduate student at Cornell University) was released. It infected 10 percent of all Internet hosts. Also in this year, *Internet Relay Chat* (*IRC*) was written by Jarkko Oikarinen.

NSF Assumes Control of the ARPANET, 1989

NSF took over control of the ARPANET in 1989. This changeover went unnoticed by nearly all users. Also, the number of hosts on the Internet exceeded the 100,000 mark.

3.3.4   1990s Telecommunications

During the 1990s, lots of commercial organizations started getting on-line. This stimulated the growth of the Internet like never before. URLs appeared in television advertisements and, for the first time, young children went on-line in significant numbers.

Graphical browsing tools were developed, and the programming language HTML allowed users all over the world to publish on what was called the World Wide Web. Millions of people went on-line to work, shop, bank, and be entertained. The Internet played a much more significant role in society, as many nontechnical users from all walks of life got involved with computers. Computer-literacy and Internet courses sprang up all over the country.

Gopher, 1991

Gopher was developed at the University of Minnesota, whose sports team's mascot is the Golden Gopher. Gopher allowed you to "go for" or fetch files on the Internet using a menu-based system. Many gophers sprang up all over the country, and all types of information could be located on gopher servers. Gopher is still available and accessible through Web browsers, but its popularity has faded; for the most part, it is only of historical interest.

World Wide Web, 1991

The *World Wide Web* (*WWW*) was created by Tim Berners-Lee at CERN (a French acronym for the European Laboratory for Particle Physics), as a simple way to publish information and make it available on the Internet.

WWW Publicly Available, 1992

The interesting nature of the Web caused it to spread, and it became available to the public in 1992. Those who first used the system were immediately impressed.

Mosaic, 1993

Mosaic, a graphical browser for the Web, was released by Marc Andreessen and several other graduate students at the University of Illinois, the location of one of NSF's super-computing centers. Sometimes you will see Mosaic referred to as *NCSA Mosaic*, where NCSA stands for the *National Center for Supercomputing Applications*. Mosaic was first released under X Windows and graphical UNIX. To paraphrase a common idiom, each person who used the system loved it and "told five friends," and Mosaic's use spread rapidly.

Netscape Communications, 1994

The company called *Netscape Communications*, formed by Marc Andreessen and Jim Clark, released *Netscape Navigator*, a Web browser that captured the imagination of everyone who used it. The number of users of this software grew at a phenomenal rate. Netscape made (and still makes) its money largely through advertising on its Web pages.

Yahoo!, 1994

Stanford graduate students David Filo and Jerry Yang developed their Internet search engine and directory called *Yahoo!*, which is now world-famous.

Java, 1995

The Internet programming environment, *Java*, was released by *Sun Microsystems, Inc.* This language, originally called *Oak*, allowed programmers to develop Web pages that were more interactive.

Microsoft Discovers the Internet, 1995

The software giant committed many of its resources to developing its browser, *Microsoft Internet Explorer*, and Internet applications.

Internet Courses Offered in Colleges, 1995

Some of the first courses about the Internet were given in 1995. Course development has been difficult, because of the rapidly changing software.

Netscape Releases Source Code, 1998

Netscape Communications released the source code for its Web browser.

3.3.5 Internet Growth

The Internet is still growing at a rate of 100 percent each year; the number of computers linked to the Internet is now 16 million and increasing. What permitted the technology to be adaptable enough to handle this amazing growth? Over the past three decades, the Internet has proven to be extremely flexible. Even though there were no personal computers, workstations, or LANs when the early Internet came into being, the emerging Internet was versatile enough to be able to incorporate these new technologies. The early researchers working on Internet technology may have had no inkling that what they were designing would one day accommodate such a thing as the World Wide Web and other applications for millions of users. The built-in flexibility has been a key to this continuing growth.

In retrospect, there were a number of key reasons for the Internet's great success:

1. Decisions were made on a technical rather than political basis, especially without the need for international political groups.

2. The Internet did not require a centralized structure that would not scale up; it was and is a distributed operation.

3. Due to the homogeneity of language and outlook, a sharp focus on the Internet itself could be maintained.

4. The Internet allowed people to do things of inherent interest, such as send and receive e-mail.

5. The software involved was free or very low cost.

Based on past history, we can assume that the Internet will continue to grow, change, and support new applications. Now, however, instead of only researchers initiating change and implementing new ideas, we also see entrepreneurs and politicians getting involved. Both small and large businesses will play an important role in setting new trends, and we can anticipate a dynamic and exciting future for the Internet.

EXERCISES 3.3 Internet History

5. Fill in five more important events in the timeline during the years 1994–97. List your sources.

6. Research one item from the Internet timeline in more detail, and prepare a one-page Web document about it. Include your sources in the document and utilize as many HTML tags as appropriate.

7. Plot the number of users of the Internet on a yearly basis since 1980. Label the x-axis with the years 1980, 1981, . . . , through the present, and the y-axis

with the number of users. Using your plot, extrapolate to predict how many users there will be one year, two years, five years, and ten years from now. What assumptions did you make in performing your extrapolation?

8. According to our timeline, in 1983 TCP/IP was mandated for all ARPANET hosts. Why was this significant in promoting the growth of the Internet?

9. Suppose you bought 100 shares of stock in each of the following: Microsoft in 1985, CISCO Systems in 1990, and Netscape Communications in 1994. Assuming you did not reinvest the dividends, how many shares of each stock would you have today? What was the original stock in each company worth, and what is their stock worth today? That is, what was your original total investment in each company, and what is your current total investment in each?

10. Look into the history of non-English Web pages. What foreign languages are supported? When did support for them originate? How widely are they supported? What do you see for the future of non-English languages on the Web?

3.4   The Way the Internet Works

Here, we present an intuitive look at how the Internet works, again using the analogy with a highway system. Keep in mind that this is a simplified description. Additional technical details can be found in the references.

3.4.1   Network Benefits

To begin our discussion of the Internet, we first identify some of the benefits of networks in general.

- *Provide convenience*—Computers on a network can back up their files over the network.

- *Allow sharing*—Networked computers can share resources, such as disks and printers.

- *Facilitate communications*—Sending and receiving e-mail, transferring files, and videoconferencing are examples of how networks promote communication.

- *Generate savings*—Networked computers can provide more computing power for less money. Several small computers connected on a network can provide as much as or more computing power than a single, large computer and will cost significantly less. Also, since resources can be shared, not everyone needs their own peripherals, which can result in a substantial cost savings.

- *Provide reliability*—If one part of a network is down, useful work may still be possible using a different network path.

- *Simplify scalability*—It is relatively easy to add more computers to an existing network.

3.4.2   Interconnected Networks and Communication

The Internet is essentially a network of networks, and its success depends upon "cooperation." Since no one person, organization, or government is responsible for the Internet, cooperation among the networks and computers that compose the Internet is paramount. This cooperation is accomplished by a common set of protocols. The protocol that determines how computers connect, send, and receive information on the Internet is Transmission Control Protocol/Internet Protocol (TCP/IP). In fact, TCP/IP consists of about 100 different protocols, and new ones are developed and added regularly. Drawing on the freeway analogy, think of these protocols as forming the "rules of the road," ranging from who has the right of way to how you register your vehicle and get a driver's license.

TCP/IP has been described as the "language of the Internet." In the same way that a common language allows people of diverse backgrounds to communicate, TCP/IP allows many different kinds of computers, from personal computers to mainframes, to exchange information. The two main protocols in the TCP/IP suite are TCP and IP. TCP permits communication between the various computers on the Internet, while IP specifies how data is routed from computer to computer.

To illustrate how TCP/IP works, consider either sending an e-mail message or requesting a Web page. In either case, the information is "formatted" according to its specific application protocol: Simple Mail Transfer Protocol (SMTP) is used for your e-mail message, and HyperText Transfer Protocol (HTTP) is used for your Web page request. Assuming that TCP/IP software is installed on your computer, the information to be sent is split into *IP packets*, called *packets* for short, and transmitted over the Internet. The advantages of packets are as follows:

- *Error recovery*—If a packet gets corrupted, only that (small) packet needs to be resent, not the entire message.

- *Load distribution*—If one area of the network is congested, packets can be rerouted to less busy areas.

- *Flexibility*—If the network experiences a failure or disruption in one locale, packets can be rerouted.

Figure 3.2 depicts the process of splitting a message into packets.

In addition to the message pieces, each packet of data also contains information about the computer that sent it, the computer it is being sent to, a

FIGURE 3.2 Schematic of Message Split into Packets and Sent Over the Internet.

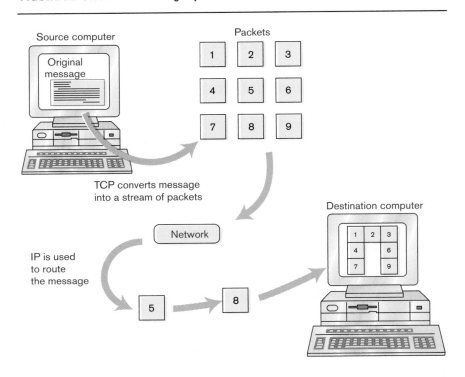

sequence number indicating where the packet fits in the overall message, and error checking information to ensure that the packet is not corrupted while in transit. The packets are reassembled after being received at the destination computer. A message is sent from the destination computer to the sending computer to resend any missing or corrupted packets. Using this method, called packet switching, it is not necessary to send the data packets in sequential order, or even over the same network route. If packets arrive out of order, the sequence numbers can be used to reconstruct the original message. Figure 3.2 shows how this is done, with packets 8 and 5 still incoming. After receiving the message, the destination computer responds, either by delivering the e-mail message to the recipient's mailbox or by servicing the request for a Web page, as required.

In the scenario we have described, suppose that a packet does get corrupted. The destination computer must send a message requesting that the packet be resent. You may wonder what happens if the "resend" message gets corrupted or lost. While we cannot get into such details here, we can say that the protocols must be complex enough to recover from all types of worst-case error situations. Protocol design is a complicated process.

3.4.3 Physical Components

In addition to the various software protocols, the Internet includes a host of physical components, as well. These components include such items as servers, routers, and the networks themselves. *Servers* are computers that answer requests for services, such as list servers, mail servers, and news servers.

A *router* is a special-purpose computer that directs data packets along the network. Routers can detect whether part of the network is down or congested and can then reroute traffic. Think of a router as a highly efficient and well-functioning traffic cop.

Networks provide the physical means to transport packets of information. The following mediums are employed:

- Copper wires, which transmit messages as electrical impulses.

- Fiber-optic cables, which use light waves to transmit messages.

- Radio waves, microwaves, infrared light, and visible light, which all carry messages through air.

3.4.4 Network Connections

Someone connecting to the Internet from home generally uses a *modem* and a regular telephone line (copper wire) to connect to an *Internet Service Provider* (*ISP*). A second modem at the ISP's end completes the connection, and the slower of the two modem speeds determines the maximum connection speed, usually 28.8 *kilobits* per second (Kbps). A kilobit is 1,000 bits. Many people use the term *bit rate* and *baud rate* interchangeably.[2] It is worth noting that some important parts of the network are still audio-based, such as the part of the phone system in your home or office. The modem is needed to convert from *analog* to digital and back again.

A business, organization, or school network typically uses *network interface cards* instead of modems to connect the personal computers that are part of their LAN. These systems often have a higher speed connection, usually 56 Kbps or better, to connect to their ISP. Such connections are usually leased from the telephone company.

Another option is an *Integrated Services Digital Network* (*ISDN*), which is slightly more expensive[3], but uses regular telephone lines and replaces modems with special adaptors up to five times faster than a traditional modem. Another possibility, not yet widely available but definitely on the horizon, is a cable television connection. Bill Gates and Craig MacCaw's company, *Teledesic*, is

---

[2] Technically, there are differences between the two.

[3] Costs vary widely across the telephone systems and are frequently tariffed to be a lot more than a modem. Installation may be hundreds of dollars and rates can be $100 to $150 per month.

putting low-earth-orbit satellites in the sky and planning to offer data service everywhere. The *Boeing Company* is also involved in this project.

If the connection to the ISP is a "driveway" in our highway analogy, then the backbones of the Internet are the "freeways." These freeways are run by *Network Service Providers* (*NSPs*). Local ISPs connect to NSP networks like IBM's Advantis or networks provided by AT&T, MCI, or Sprint. The connection between the ISPs and NSPs is over leased-lines from local telephone companies. These phone lines typically transmit data at a rate of 1.54 *megabits* per second (Mbps). A megabit is 1,000,000 bits.

The NSPs lease or buy lines consisting of copper wire, fiber-optic cable, or satellite communications from such companies as AT&T, MCI, Sprint, or LDDS WorldCom. The NSP networks, like a freeway, can operate at very high speeds and can transmit a lot of data over long distances.

To summarize, to access the Internet, a user connects to a local ISP through a modem or ISDN adapter (or possibly some other method). The ISPs connect to the larger NSP networks through leased-lines from the local telephone companies. To transmit a message over the Internet, TCP/IP divides the message into packets that are sent over the lines and directed via routers to their destination. When the packets arrive at their destination computer, they are reassembled and the destination computer responds to any request. Figure 3.3 illustrates the entire process.

3.4.5   Client-Server Model

In Figure 1.8, we showed how mail clients and servers interact. Browsers, also called Web clients, and Web servers were described in Chapter 2. The interaction between clients and servers can be generalized to other applications. Figure 3.4 depicts the general arrangement known as the *client-server model*. In the picture, only one client is shown, although a large number of clients typically use a small number of servers. A client makes a request to the server, and the server responds by satisfying the client's request.

The client-server model provides many of the network benefits described in Section 3.4.1. For example, the client-server model is easily extendable and therefore scales well; that is, new clients and servers can be added incrementally as more users come on-line and the demand for services increases. Many clients can share the resources provided by a single server. This eliminates the need for each client to have their own "copy" of those resources.

Each Internet service has its own associated set of clients and servers. We have already seen examples in the e-mail and Web domains, and later we will see others.

3.4.6   IP Addresses

Each computer and router on the Internet must have a name so that it can be uniquely identified. After all, how can an e-mail message be delivered if

FIGURE 3.3 Illustration of a Message Split into Packets, Routed to Destination through an ISP and an NSP, and Reassembled.

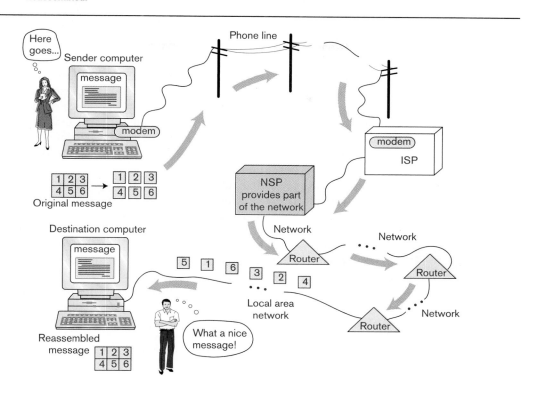

FIGURE 3.4 Illustration of the Client-Sever Model.

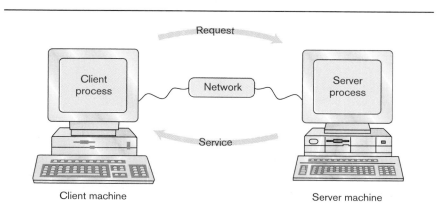

there is an ambiguity in its destination? While the *domain name* provides a convenient people-oriented computer naming framework that uses symbolic names, computers are better suited to manipulating numbers. *IP addresses* are numerical names that uniquely identify each computer on the Internet.

An IP address consists of 32 bits, or four *bytes*. (A byte consists of eight bits.) In Table 2.2, we saw that the largest possible eight-bit number was 255. Thus, one byte can represent a number from 0 (00000000) to 255 (11111111). Oversimplifying, each IP address consists of a *network component* and a *host component*. Figure 3.5 part (a) illustrates the concept. Each of the four bytes of an IP address can represent a natural number from 0–255. It is common to express IP addresses as four natural numbers separated by dots. Figure 3.5 part (b) shows an example, and part (c) shows the corresponding binary numbers.

IP addresses play a vital role in the routing of packets over the Internet. Source and destination IP addresses are included in each packet. In essence, the addresses provide directions on where the packet should go. How are IP addresses assigned? A central authority manages IP addresses; otherwise, conflicts might arise. The *Network Information Center* (*NIC*) is in charge of assigning IP addresses. (However, there is a plan to expand to multiple registries. It is not yet clear how they will coordinate.)

What is the relationship between IP addresses and domain names? IP addresses are 32-bit numbers, whereas domain names are easy-to-remember symbolic strings. When you type in an e-mail address, you always enter a symbolic string such as

rudolph@northpole.org

FIGURE 3.5 Part (a), Schematic of an IP Address; Part (b), the Customary Way of Writing an IP Address; Part (c), the Binary Equivalent of the IP Address from Part (b).

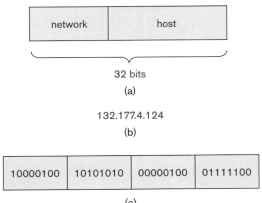

How does a computer use this, since it needs to work with IP addresses? A program called a *resolver* takes care of the translation; that is, the program converts a symbolic name into its corresponding IP address. Think of the resolver as functioning in a manner similar to telephone directory assistance. On some systems, there is a program that allows you to enter an IP address and obtain its symbolic name back, or vice versa. An example on UNIX-based systems is the program *nslookup*. It is important to realize that for each symbolic name, there is a unique IP address.

On occasion, you may come across an IP address. For example, while surfing the Web you might see a message such as "invalid IP address." When this happens, there is usually a problem with the computer name you specified. Either you typed the name wrong, or the server was not responding for some reason. Other times, you may see a message like "host name could not be resolved." Again, the first thing to do is check your spelling. If you configure TCP/IP or *Point-to-Point Protocol* (*PPP*) for your computer, you may have to enter an IP address for your domain name server. If you use an ISP, you may need to obtain this address from them. (See Appendix A for more about ISPs.)

A permanently assigned IP address, one that is given to a computer or router connected to the Internet, is called a *static IP address*. If you connect to the Internet through an ISP, then typically each time you connect, you will be assigned a different IP address, called a *dynamic IP address*, from the ISP's pool of IP addresses.

3.4.7   Internet Protocol Version 6 (IPv6)

IPv6 is the latest version of the IP routing protocol. It is still being developed. The new protocol is necessary to accommodate the greater demands being placed on the Internet. The major changes in the new version will be:

- *More addresses*—This will be done by increasing the IP address size from 32 to 128 bits.

- *Simplified IP headers*—The number of header fields needed in an IP packet will be reduced.

- *Added security features*—The new protocol will provide greater support for privacy and security.

In other words, IPv6 is being designed with many efficiency considerations in mind. You can find additional details about IPv6 on the Web.

3.4.8   Web Page Retrieval

Suppose you select the following URL:

```
http://pubpages.oklahoma.edu/~bosworth/food/menu.html
```

either by clicking on a hyperlink or by entering this URL in the location field of your browser and hitting the return key. How is this Web page retrieved? Having discussed browsers, DNS, IP, IP addresses, TCP, URLs, and Web servers, we are now in a position to answer this question. The following steps occur, in the order listed:

1. Based on your actions, the browser determines that the URL you selected was

   `http://pubpages.oklahoma.edu/~bosworth/food/menu.html`

   The how part of the URL is `http`.
2. Utilizing the where part of the URL, the browser queries the DNS for the IP address, `pubpages.oklahoma.edu`.
3. In this case, suppose the DNS responds with the IP address

   `172.177.173.2.`
4. The browser next establishes a TCP connection to `172.177.173.2`. The default *port* for Web servers is port 80. Think of the browser as letting `pubpages` know it wants to talk on a reserved category called "80."
5. The browser then sends a message asking for the what part of the URL,

   `~bosworth/food/menu.html`

6. The server, `pubpages.oklahoma.edu`, services this request and sends back the file `menu.html`.
7. The TCP connection is closed. This ends the "conversation" between the client and the server.
8. The browser renders the text portion of the HTML code contained in the file `menu.html`.
9. The browser repeats these steps to obtain any in-line images contained in the file `menu.html`.
10. The browser displays the images as they are retrieved.

This procedure can easily be generalized to illustrate how any Web page is retrieved. Notice that a separate TCP connection is required to bring over each in-line image. This fact explains why images load last when a new Web page is brought over, and why a Web page containing a lot of images takes a significant amount of time to load.

EXERCISES 3.4 The Way the Internet Works

11. Find out what ISPs are available in your area and what connection options they offer. Can you directly connect to an NSP? What is the highest (fastest) connection rate available to you?

12. Determine how your school, organization, or company is actually connected to the Internet. Draw a small map to illustrate your connection.

13. Write a short Web page about the current state of cable television connections. List your sources.

14. Define and explain two additional positive benefits of networks.

15. To help you appreciate the complexity of protocols, describe in detail the "protocol" for making a telephone call. For example, you lift the receiver off the hook, . . .

16. In Section 3.4.5, we explained how the client-server model supports the network benefits of scalability and sharing. How does the client-server model relate to the other network benefits described in Section 3.4.1?

17. There is a movement to change the addressing scheme of the Internet. Conduct a search on-line for proposals about new addressing schemes and summarize your findings.

18. Experiment with your browser to try and find the answer to the following question: If an HTML document contains a number of identical copies of the same image, does the browser bring over multiple copies of the image, or is it "smart" enough to reuse the first copy it brings over?

3.5 Internet Congestion

The Internet functions amazingly well for such a heavily used system. However, the number of users and their demands continue to grow almost without bound. In this section, you will learn what researchers are trying to do to reduce congestion on the Internet. First, we consider some of the limiting factors, from a user's point of view.

Once you get a network connection (assuming you have a modem with a speed of 28.8 Kbps or higher), the factor limiting how quickly you can view Web pages often becomes the speed with which your computer renders the pages. The computer speed depends on a complex balance of CPU speed, bus speed, memory quantity, disk speed, and so on. In addition, there is a hierarchy of link speeds, with major backbones aiming for OC12 (622.1 Mbps),[4] more and more regional and site-direct T3 (44.7 Mbps) links, and T1 (1.5 Mbps) links to institutions. The slowest link speed involved in a connection determines the overall level of performance. The rate of growth of the Internet is so rapid that it is hard for technological improvements to keep up.

3.5.1 World Wide Wait Problem

The phrase *World Wide Wait* has been around for awhile, especially overseas, where connections are notoriously slow. For example, in Spain, Internet users

---

[4] OC stands for *optical carrier*.

say,

```
Espera en la Red Mundial
```

The literal translation of this phrase is "Wait in the World Network." It refers to the ever increasing delays experienced when trying to access information on the Internet.

   With the advent of the World Wide Web and the development of graphical browsers came a surge of interest in the Internet. This increase in the number of Internet users, coupled with the accompanying requests for Web pages containing elaborate in-line images, sound, and video clips, has degraded the speed of the Internet to the point where the Information Superhighway sometimes appears to have a traffic jam. Although new technologies are being employed to remedy the situation, the problem persists and is getting worse. We will discuss what is currently being done to address the problem and what we can expect in the future.

3.5.2   Technical Solutions

Researchers working in conjunction with the *World Wide Web Consortium*[5] (*W3C*) are addressing the issue of network congestion. One of their stated goals is to "save the Internet from the Web" by developing new technologies to help relieve the slowdown that has resulted from retrieving and displaying Web pages.

   Some of the solutions offered involve HTTP itself, as well as improvements in the way HTTP and TCP/IP interact. In particular, the researchers have focused on the following issues:

- Improving the process of connecting to a Web server.

- Introducing new techniques to expedite Web page requests.

- Refining how a URL is resolved, using "persistent connections" that make it more efficient to retrieve pages from the same Web server.

   W3C researchers have also put forth some suggestions for Web page design. Since Web page content (that is, the graphics, sound, text, and/or video that make up the page) dictates download time, one recommendation is to avoid unnecessary graphics. *Cascading Style Sheets* (*CSS*), a Web page design tool, also has the potential to improve download time. Finally, the researchers recommend using the *Portable Network Graphics* (*PNG*) format over the *Graphics*

---

[5] The W3C provides an open forum to facilitate communication between individuals dealing with matters related to the World Wide Web. The *Internet Engineering Task Force* (*IETF*) plays a similar role regarding the Internet. The official documents developed by the IETF are called *Requests for Comments* (*RFCs*).

Interchange Format (*GIF*) for images on Web pages, since PNG images are generally smaller than GIF images and they render more quickly.

Another active step being taken is a reduction in the size of router tables by rearranging how blocks of addresses are identified. Routers face a formidable task when data flows at 44.7 Mbps or faster. They must examine each packet to see where it is going, look up that destination, and then send the packet on its way. They cannot fall behind, because they would never catch up. Packets that cannot be resolved in the *threshold time* are thrown away and must be retransmitted.

By developing these and other technological improvements, researchers and the W3C are attempting to ease the Internet congestion created by the World Wide Web. It is believed that these new technologies can reduce Internet traffic due to the Web by up to 50 percent. However, network traffic is increasing in many dimensions. New users are being added, and experienced users are requesting more information and spending more time on the Web. Thus, although suggestions like those given here are worthwhile if many users treat them seriously, they are actually expected to provide very little noticeable relief.

3.5.3 Issues and Predictions

While recent technological advances have been significant, they have not yet managed to alleviate the World Wide Wait problem. Some users attempt to deal with the slowdown by using the Internet during less busy periods. Still others, frustrated by the delays and failures in transmitting and receiving information, may severely limit their use of the Internet. Plans have moved forward to create the "Internet II" to be used exclusively by academia and researchers, with no commercial traffic permitted, and with operation at much higher speeds. A number of businesses are already bypassing the Internet and are creating isolated *intranets* for their companies. An intranet is a private network. Such networks can have their own internal Web.

One reason why the Internet has become so popular, especially in the United States, is that it is very inexpensive. Most ISPs offer a flat-rate plan that allows for unlimited usage for a single, low monthly fee. Some economists feel it may be time to charge more for the use of the Internet in order to limit demand.

Originally, the government financed the Internet. Now, users pay ISPs to connect them, ISPs pay NSPs, and NSPs pay the long-haul carrier. The payments are not based on hours of use, but on how much capacity is provided. This contrasts with the telephone billing system, in which there is a unit charge based on how far the call travels and/or how long it lasts. If the telephone system were run like the Internet (and were just as cheap), the demand for phone service would greatly increase, as would delays and busy signals. However, since the telephone system is comparatively so expensive, people have figured out ways of having "phone" conversations over the Internet.

To reflect the true costs involved in using the Internet, some economists have suggested prioritizing information and then charging more for high-priority transmissions. Another idea is to charge (more) for transmissions that occur when Internet traffic is heavy. How to meter usage once a billing method is selected is currently being investigated.

It seems clear that unless technology offers a viable solution to the World Wide Wait problem, our days of surfing the Web for "free" could be numbered.

EXERCISES 3.5 Internet Congestion

19. How long have you been using the Internet? Do you notice a more significant delay in retrieving documents now than you used to? Do you try to restrict your access to the Internet to certain times of the day? Have you ever surfed the Web from a foreign country? Describe your experience.

20. How many times a week do you get a "busy signal" when you try to connect to a site? That is, the server refuses your connection. Do you use your browser's **Stop** button more frequently now than previously to cancel a Web page request that seems to be taking forever to load?

21. A lot of people in Europe and Japan are heavy users of the Internet. Based on the different time zones and the fact that the majority of Internet users are located in the United States, when would you expect the World Wide Wait problem to be the worst? Explain your answer.

22. Write a short Web page describing the current state of the art for placing a "phone" call over the Web. Describe the hardware necessary and its cost. Do you believe this technology will take a significant amount of business away from the telephone companies?

23. What methods are being suggested to meter network usage? Describe them. In a Web page, summarize your thoughts about metering network usage.

3.6 Internet Culture

An entire culture has sprung up around the Internet. What began as an exclusive club for researchers and academics is now open to the masses. Some of the original club members are not happy about this transition. The Internet has emerged from being a research medium to one that includes advertising, commerce, and forums for exchanging ideas on a near infinite set of subjects. Here we describe the philosophy of this unique culture and some important issues to bear in mind while browsing.

3.6.1 Critical Evaluation of Information

Since the Internet is not regulated for content (and there are no immediate plans to regulate it), anything and everything can be found on the Web. The editorial control that is applied to traditional print media is missing. Being able to distinguish between inaccurate and accurate information is a necessary part of the Internet culture. Experienced Internet users know that not everything published on the Web is sound. They are cautious about believing anything they read, looking at information with a critical eye.

To find valuable information, you need to be able to sift through Web pages and separate the useful from the useless, the valid from the invalid. Suspect information can appear on the Internet in any form: e-mail messages, mailing lists, newsgroups, or Web pages. For example, in the 1996 U. S. presidential campaign, bogus presidential Web pages were published.

What are some reliable forms of information? Web presentations that contain refereed and reviewed information, or are monitored for accuracy, could be considered reliable. An example would be electronic journals whose contents are refereed. Commercial presentations also try to provide accurate and up-to-date information, since their reputations depend on presenting valid information. Some authors, by the very nature of who they are, can be trusted to display only accurate information. For example, if Miguel Indurain, five-time winner of the Tour de France, had a Web presentation about cycling, you could assume that the information about the Tour de France would be accurate.

How can you critically evaluate information? You might start by asking the following questions:

- *Who wrote the information?*—Was the person who wrote the material knowledgeable and careful? Was the author aware of what others have written? Does the author have a reputation to uphold? Can the author be trusted? What is the author's professional background?

- *Is the writing quality high?*—A document riddled with typos is more likely to have inaccurate content than a carefully crafted Web page.

- *Is the document up to date?*—Try to determine whether the information is current. When was it last updated? Does the document deal with up-to-date information?

- *Are there obvious errors in the content?*—For example, if you know that the game of baseball requires 11 players on a team and the document you are reading says it only requires 6, be wary.

3.6.2 Freedom of Expression

The lack of regulation that permits the proliferation of information on the Internet also facilitates the interchange of ideas. Anyone with an Internet connection can express their views globally. This also allows small groups of people with something in common to find each other. For example, there are people with very rare medical problems who can offer each other support and can exchange experiences in coping. Many believe that this freedom of expression is the best, and most defining, feature of the Internet. A large number of Web authors exhibit this fact by displaying a small blue-ribbon graphic at the bottom of their pages in support of on-line "freedom of speech."

Related to the idea of personal expression is another aspect of Internet culture—not everyone agrees that everything and anything should be publishable. For example, some people find the availability of obscene or offensive material on the Internet unacceptable. Other people worry that small children may stumble across something they should not see or read. These concerns are definitely valid and several camps are busy discussing them.

What constitutes obscene material? What is pornographic material? Certainly, not everyone is in agreement on the answers to these questions. However, there are legal definitions of *obscenity* and *pornography*, and corresponding bodies of case law. Problems arise when there is an attempt to legislate restrictions without careful definition of what is covered. For example, child pornography is subject to proscription by special laws and that might seem clear—but questions are now rising as to whether they apply to computer animations or simulations in which no real children are involved.

In the early days of the Internet, many users were young, technically oriented males from universities and research laboratories who were, relatively speaking, like-minded. Today, people from all walks of life are getting information from and publishing on the Internet—children as well as adults. For some people, the accessibility of offensive material has created a negative image of the Internet. Unfortunately, some people lose sight of all of the great benefits that the Internet provides:

- More educational opportunities for both children and adults.

- The ability to communicate more readily with others all over the world.

- The sharing of research ideas and information.

- The convenience of performing many functions, such as banking and shopping, on-line.

- Opportunities for entertainment.

- Rapid and global dissemination of important information.

- Worldwide discussion forums to promote solutions to global problems.

Prohibiting material on a specific topic from being published on-line diminishes freedom of expression. On the other hand, parents, for example, may feel their freedoms are being violated if they cannot have the Internet and its benefits without risking unintended exposure of their children to "un-acceptable" information. The issue of *censorship* is a volatile one that has both supporters and opponents in the Internet community, and the issues involved in worldwide censorship are very complex.

One possible compromise being tried is the use of labels to identify the content of Web pages. Authors can thus steer viewers away who might be offended by the material or for whom the material might be inappropriate. The *Platform for Internet Content Selection (PICS)* provides a set of technical specifications for designating such labels. (PICS was initially designed to aid parents and teachers in limiting what children could access on the Internet.) The PICS specifications work with vendor-supplied filtering software (some of which is already built into browsers) and rating services. Some currently available PICS-based software systems and their distributors are: *Net Shepherd* (Net Shepherd), *Cyber Patrol* and *Cyber Sentry* (Microsystems Software), *Surfwatch* (Spyglass), and *Internet Explorer 3.0* (Microsoft). In addition, some ISPs offer PICS-based filtering.

The Internet leaps borders and boundaries, so that content that is legal in the jurisdiction where it is created could be illegal in some other part of the world. Does this mean that the creator should be subject to the lowest common denominator of "acceptable" conduct? Experience with book publishing shows that there are few, if any, books that would not be subject to banning by some group, if they could. This suggests that the proper place for content control is with the content viewer, not the creator. PICS is one approach to this solution.

Although PICS provides an alternative to government control of Internet content, many are still concerned about promoting any censorship technology. The case concerning filters and PICS is not clear cut. The issues involved are similar to those concerning media, such as cassette tapes and compact disks. However, although you can buy a CD without buying an offensive one, you cannot get only partial Internet access without using filters. There are many overlapping and conflicting interests in the way *cyberspace* develops, and we all need to be active, informed participants.

3.6.3   Communication Mechanisms

Another aspect of Internet culture is created by the communication channels that the Internet has spawned. People from all over the world are able to exchange ideas via e-mail, *Internet Relay Chat (IRC)*, mailing lists, newsgroups, and so on. Since there are no facial expressions, voice inflections, or body language to convey or interpret these communications, users must avoid ambiguity or misunderstanding by either spelling things out completely or using *emoticons*. Table 3.1 depicts a number of common emoticons. While *videoconferencing* is a way to include the otherwise missing audio and video, this technology is still in its infancy.

TABLE 3.1 Emoticons.

Emoticon	Meaning
0:-)	angel
>:-)	devil
:-(	frown
:-[	grim
:-D	laugh
:-)	smile
:-o	surprise
;-)	wink

To save time when typing messages, users sometimes employ a (friendly) shorthand for commonly used phrases:

- AFAIK—As Far As I Know.
- FOAF—Friend Of A Friend.
- FWIW—For What It's Worth.
- IMHO—In My Humble Opinion (usually not so humble, of course).
- ROTFL—Rolling On The Floor Laughing.
- RTFM—Read The $#*&ing Manual (you might see this if you ask for something that is explained in a manual).
- YMMV—Your Mileage May Vary.

Even when you are trying to be very clear (or perhaps because of it), people may sometimes misinterpret or take offense at a message. Being rude or overly confrontational is called *flaming*, and such messages are called *flames*. Some people find it easy to be rude when they do not have to confront a person face to face, while others are just plain unaware of how their messages come across. Flaming is not considered appropriate on the Internet; it violates the commonly accepted guidelines of netiquette.

3.6.4 Advertising

Prior to 1995, there was very little advertising on the Internet. Along with the Web has come an avalanche of advertisements. Users who began surfing the Web in 1996 or later are accustomed to ads, but those who started earlier remember the days of very limited advertising.

Ads generate huge incomes for such companies as Netscape Communications, Infoseek, and Yahoo!. The Web pages of these companies get millions of hits per day, so an ad placed on one of their Web pages has a tremendous audience. Naturally, marketing experts take advantage of this potential consumer base.

Most of the ads shown on Web pages are clickable images. They are typically about one inch high and three to four inches wide. Many of the most popular Web pages have *revolving advertisements*; that is, each time you revisit the page, or while you are visiting the page, you get a different ad. The ads usually consist of a carefully designed graphic with a catchy phrase superimposed over it. When you click on the graphic, the advertiser's Web page loads.

Many users manage to browse the Web without paying too much attention to the advertisements, other than noticing that the ads slow down the loading process. Obviously, the ads influence some people, because companies continue to invest huge amounts of money in them.

The marketing techniques for advertising on the Web are also becoming more sophisticated. An industry is developing to monitor who visits what sites, so that ads can be targeted more specifically to certain users.

The style, form, and content of ads is a rapidly emerging part of the Internet culture.

3.6.5 Societal Impact

The Internet has had an enormous impact on society, and its influence will no doubt continue. Nearly all facets of life have been affected. Many people now work in Internet-related jobs, either building computer network components, writing software, creating Web pages, performing marketing research, designing graphics, or conducting business on the Web. Unless you are hiking the Pacific Crest Trail, you will probably run across at least a few URLs everyday (yet even on the trail, people are carrying laptop computers). Many people obtain all their information and perform most of their communication using the Internet. Such things as weather, news, stock prices, and travel information are accessed by millions of users every day. It is difficult to think of any areas of society that have not been strongly impacted by the Internet.

EXERCISES 3.6 Internet Culture

24. In a single Web page, summarize your thoughts about whether or not the Internet will ever be regulated; that is, will the contents placed on the Web ever truly be limited and controlled?

25. Locate two pages on the Web with the following properties: one of the pages should contain what you believe to be inaccurate information, and the other should contain what you perceive as accurate information. Compare and contrast their differences in a Web page you develop.

26. What is your personal opinion about the issue of censorship on the Internet? Do you think filters are a viable solution?

27. Decipher the following shorthand expressions: `BTW`, `FYI`, `KSA`, and `WRT`. List five other common abbreviations you have encountered, along with their meanings.

28. How much money does a company pay to place an ad on Netscape's Web pages? List your source(s).

29. Describe how ads affect you during Web browsing. What do you think is the future of Web advertising?

30. How has the Internet affected your education?

31. What role do you see the Internet playing in the future of society? Describe the facet of society on which you feel the Internet has had the greatest impact. How about the least impact?

3.7   Business Culture and the Internet

Many businesses are recognizing opportunities in on-line activities. In newspapers and magazines, we constantly see URLs listed. For example, the following URLs may look familiar:

`www.mastercard.com`	Mastercard International
`www.atlanticrecords.com`	Atlantic Records
`www.toyota.com`	Toyota USA

Even on television,[6] we are flooded with URLs, such as:

`www.sixflags.com`	Six Flags Theme Parks
`www.edf.org`	Environmental Defense Fund
`www.dell.com`	Dell Computer Corporation

Not too long ago, there was a big debate among advertisers about whether or not to include the `"http://"` part of a URL in a televised ad. On the radio, you will hear announcers giving out URLs almost constantly. Some radio announcers still do not know how to read a URL; nevertheless, they continue to try, and most are getting the hang of it.

 Another question that many users and companies pose is, "Is it safe to do business on the Internet?" Several large computer companies have television

---

[6] Compare the attention span of someone channel surfing versus surfing the Web. The main difference is that instead of using your remote to view 57 channels, you use a mouse and have millions of channels from which to choose.

ads trying to convince you that it is safe. Consumers are worried that businesses are going to "find out everything about them." Businesses are concerned, because they are not knowledgeable enough about the Internet.

Because business on the Web is still in its infancy, many questions remain unanswered. Here, we explore business issues and look at several businesses that are successfully using the Internet to increase their markets.

3.7.1 On-line Businesses

The Internet functions nicely as a means of facilitating business communications within a given company, as well as between companies. The Internet is also an excellent venue for advertising and conducting trade with consumers. It is currently possible to shop for goods and services through on-line catalogs; subscribe to on-line versions of magazines and newspapers; and purchase software. These are just a few types of business transactions taking place on the on-line marketplace.

In addition to lowering transaction costs, the Internet is transforming the marketplace into a global environment in which businesses and consumers are no longer restricted by their geographical locations. For companies, this means more potential customers; for consumers, this means a greater selection of services and products. This revolution is literally changing the way a lot of companies do business. Here are a few of the new and interesting business models on the Internet.

* Advertising
    * Example: *AltaVista*.
    * AltaVista is a *search engine*.
    * Advertisers pay for the search service, and consumers can be targeted for specific types of ads on the basis of their search requests. This specialized type of advertising is very effective at reaching target markets.
* Marketing
    * Example: *AmericaNet*.
    * This service helps businesses get started on the Web and also has a section for classified ads.
    * Businesses can either purchase Web presentations or advertise through AmericaNet.
* Partnership
    * Example: *FedEx BusinessLink*.
    * With FedEx BusinessLink, customers can buy goods and services on-line, and can have them delivered through a joint venture between the merchant and Federal Express (FedEx).

- Retail
  - Example: *L. L. Bean*.
  - L. L. Bean sells outdoor gear and clothing.
  - Consumers can view, select, order, and pay for their merchandise on-line.
- Service
  - Example: *Internet Travel Agency, Inc.*
  - This was one of the first full-service on-line travel agencies (great for comparing airline fares).
  - Travelers can examine, reserve, and pay for their tickets on-line.
- Software
  - Example: *Netscape*.
  - Buyers are able to download software, use it, and pay for it on-line. One of the key features here is that potential buyers can try a product out for a month or so before purchasing it.
- Subscription
  - Example: *The Wall Street Journal Interactive Edition*.
  - This is an electronic newspaper. Some features are available to non-subscribers.
  - Subscribers can view a continuously updated version of the newspaper on-line 24 hours a day. This is the fastest method for getting the most current news. Since many business investments are time-critical, having the latest available information can be paramount.

Notice that in each situation, business costs are reduced by utilizing the Internet, and the convenience afforded the customer over the traditional method of conducting business is significant. This combination will lead to many more successful businesses and satisfied customers.

3.7.2 Three Sample Companies

FedEx BusinessLink

The partnership business model deserves special mention, as this "virtual enterprise suite" strategy is a new concept. Federal Express provides FedEx BusinessLink, a service that creates a partnership between FedEx and companies seeking to tap the on-line market. By providing a method for a business to offer on-line ordering of their products and then delivery of the order anywhere in the world, FedEx BusinessLink makes it possible for any size of company, located anywhere in the world, to expand their market without additional overhead and without significant risk. For the customer, this model means more convenience; the selection is great and the delivery is fast.

FedEx BusinessLink works by supplying software to companies so they can create and maintain their own on-line catalogs. The Web catalogs reside on a *secure server* at FedEx. Customers order through the on-line catalog, and a customer confirmation number is tied to a FedEx tracking number. FedEx BusinessLink forwards the information to the retailer electronically, and the retailer fills the order. FedEx takes care of shipping the order. At any point, from pick-up to delivery, both the buyer and the retailer can track the shipment on-line. This makes the system very convenient.

food-online.com

Another interesting on-line business is grocery shopping. Shoppers visit a grocery store Web presentation and select items to purchase, filling their virtual shopping carts. One company providing support for on-line grocery stores is food-online.com. This company creates and maintains Web presentations on the Internet for various grocery retailers. Instead of employing proprietary software on private networks, food-online.com takes advantage of Internet technology by using the Netscape browser to display the virtual grocery store.

Using point-and-click tools, shoppers are able to select items (and put them back), view pictures of them, read nutritional information, determine the sizes available, and price the goods—in short, everything but actually handle the item. At "check-out," the order is totaled, and customers can select their method of payment, as well as their pick-up or delivery preference. In addition, the store can remember a customer's food preferences. In this way, the consumer can begin with a partially filled "cart."

Using this business model, the grocery retailer is able to update product prices in real-time, plus utilize a warehouse-based operation, (such as *Groceries to Go*), if desired. Keeping track of inventory can also be simplified. Obviously, the convenience to the customer is great.

ONSALE, Inc.

Another unusual on-line business application that has proven highly successful is ONSALE, Inc. (ONSALE, for short). This is an on-line interactive real-time *auction* that markets refurbished and closed-out electronic equipment. There are many other forms of on-line auctions, as well. With ONSALE, auctions are held on-line on Mondays, Wednesdays, and Fridays. Participants browse through available items, bidding and rebidding, until the bidding concludes. If you are outbid, you will receive an e-mail message so indicating. Winning bidders are notified by e-mail, and ONSALE transmits the orders to the participating merchants, who fill the orders and collect the payments. ONSALE has been rated as a top shopping business, because it not only provides some real bargains, it is also extremely entertaining.

3.7.3   On-line Business Hurdles

Thus far, we have mentioned only three of the many companies conducting business on-line. The concept sounds great, and many businesses are turning a handsome profit. However, some hurdles must be overcome. Here we discuss some of the problems and concerns of on-line buyers and sellers.

Probably the most significant consumer concerns about doing business on-line are *privacy* and *security*. This is a complex subject, but it is appropriate to include some remarks here. When disclosing personal information and revealing spending habits on-line, consumers want assurance that the information will go no further. It may be a bit unsettling to revisit a Web page and have them ask if you would like to pick up where you left off. Some users could also get nervous if their favorite on-line catalog remembers their hat sizes, shoe sizes, and credit card numbers. What is to prevent this information from falling into the wrong hands? These data are actually stored on your hard disk in a file usually called `cookies`. They may also be stored in a `cookie` directory. For our discussion, we assume that the personal data is stored in a file called `cookies`, and we refer to each entry in the file as a *cookie*.

3.7.4   Cookies

Sometimes when you visit a Web page, information about you is collected. It might be your name, password (if you are registering for that page), preferences, your computer's name, flags that keep track of what you looked at, credit card number, phone number, address, etc. We often volunteer this information by filling out a *form* on the Web page. Or the information exchange may be inherent in the transaction process, such as supplying a mailing address when ordering. Parts of this information may even be *encrypted*.

A Web server sends this information to your browser, and the data is written to the `cookies` file stored on your disk. This process is known as *setting a cookie*. Some browsers allow you to select an option that notifies you when a cookie is to be written; you have the option of not allowing the cookie to be written. Using the `cookies` file, a Web server can keep track of the Web pages you visit. The next time you visit a particular Web page, the server will search the `cookies` file, retrieve the information stored there, and use that information to customize its Web page to accommodate you.

In actuality, the amount of data that can be stored in a cookie is very limited. The most likely scenario is to store an `id` for you, fetch that `id` from the cookie, and then look you up in the server database for your more detailed profile and history.

The purpose of putting information in the `cookies` file on your disk is to reduce the server's search time in locating a specific cookie, namely yours. Since the `cookies` file is limited in size (it typically contains only about 300 cookies), locating a specific cookie can be fast. However, the size limitation also means that after a period of time, some cookies must be removed. The least recently used entries are deleted when space is needed. Cookies also specify an expiration date after which time they may be removed from the file. The term *persistent cookie* derives from the sometimes long periods of time that elapse before entries are deleted from the `cookies` file.

One concern about the `cookies` file is that information may be retrieved and used to determine an individual's personal habits. Aside from credit card account numbers, it is generally felt that the recorded information is fairly harmless. Credit card security is a valid concern, but such numbers are encrypted. Many feel that the pluses of cookies outweigh the potential negatives. For instance, cookies are used to keep track of items on your shopping list as you go through an on-line catalog. When using the Netscape browser's preference options, your preference selections are saved as cookies on your disk, so your preferences do not have to be set every time you start the browser. These are examples of two interesting and harmless uses of cookies.

Consumers will have to decide on the cookie controversy on their own. The general feeling is that their benefits and functionality are sufficient to justify their use. And, as with the use of credit card transactions, there is a tendency for journalists to stress the possibility of risk.

3.7.5 Business and Safety/Security on the Web

Probably the biggest concern of consumers conducting business on-line is the issue of secure payment: Is it safe to use your credit card on-line? The more relevant question may be whether it is as safe to do business on-line as it is to conduct business in other ways. If conducting business on-line is no more vulnerable than conducting business over the telephone, then many users will feel comfortable about on-line purchasing.

While a concern for and understanding of the transaction process is healthy, the news media tend to exaggerate the risks. In reality, even the normal use of credit cards is not without similar risks. When ordering by credit card over the telephone, we trust the retailer to handle our information with care and confidentiality. In a restaurant, if we pay with a credit card, we assume the establishment will not use our card for any other purpose.

Mechanisms for ensuring secure payments are currently being developed in the private sector. *Secure Electronic Transactions* (*SET*) is a new technical standard to be implemented by Visa and MasterCard to make credit card payments over the Internet more secure. Other payment options being developed include *electronic money*. The trend is that business transactions over the Internet are becoming more widespread and also more secure. It is only a matter of time before the relative level of security matches that of other transaction mechanisms.

3.7.6 Legal Environment

While concerns about secure payments may scare some potential on-line consumers, issues concerning the legal implications of doing business on-line have discouraged some companies from taking their businesses on-line. Consumers and businesses recognize that electronic commerce is new and uncharted territory. Without a predictable legal structure and without a guarantee that governments will not suddenly impose taxes and tariffs on trade conducted over the Internet, a number of companies find the risk too overwhelming. However, this number is probably small compared to the pragmatic and legitimate concerns that Web page visits will not translate into sales.

3.7.7   U. S. Government's Commitment to Electronic Commerce

In July 1997, President Bill Clinton made a strong commitment to promote global electronic commerce with the release of the report "A Framework for Global Electronic Commerce," or *Framework* for short. The Framework defines how policy on the *Global Information Infrastructure* (*GII*) should be developed to promote "the development of a free and open global electronic marketplace." The report is significant both for what it plans to do and for what it will not do. The underlying principles, summarized from the Framework, are as follows:

1. Governments should encourage self-regulation of the Internet and encourage the private sector to take the lead in organizing standards when needed.

2. Because technology is changing so quickly, governments should not attempt to regulate or restrict electronic commerce on the Internet, since such policies may be obsolete before they are even enacted.[7]

3. Governments should provide a legal environment to support electronic commerce and to protect consumers when necessary.

4. Governments should acknowledge the uniqueness of the Internet by not trying to impose other regulatory structures on it, such as those applying to the telecommunications industry, radio, and television.

5. Electronic commerce on the Internet should be promoted globally in a consistent manner, regardless of where the buyer and seller reside.

The report addresses financial, legal, and market-access issues and advocates a "hands off" policy whenever possible. The recommendations made cover the government actions that may be necessary to ensure secure electronic payment systems and to safeguard personal privacy.

Currently, the Internet and electronic commerce seem poised for a free-for-all without any clearly defined boundaries. As governments around the world formulate their policies regarding electronic commerce, and the

---

[7] Except for the U. S. policy on the export of encryption technology.

legal environment becomes better defined, we anticipate that both consumers and businesses will grow to feel even more comfortable conducting business on-line.

EXERCISES 3.7 Business Culture and the Internet

32. Record five URLs that you encountered either in off-line magazines, newspapers, radio, or television. Note the media form in which the URL was found.

33. Research an on-line business. Write a short summary of how the company functions.

34. Do you feel cookies are an invasion of a user's privacy?

35. Copy an entry from your `cookies` file and label what each part represents.

36. Has anyone you know purchased something on-line using a credit card? Were there any problems? Have you done so? If so, describe your experience.

37. What are your thoughts about the role the U. S. government is taking towards commerce on the Internet?

3.8 Collaborative Computing and the Internet

Collaborative computing is currently generating great interest in many different areas of computing. Here, we will describe what collaborative computing means, explore several examples, and consider its impact. We will also examine where collaborative computing is going.

3.8.1 Collaborative Computing Defined

Collaborative computing is defined as applications that allow the sharing of information and resources between two or more people. The World Wide Web, with its panoply of Web pages, is a collaborative computing platform that employs HTML and Web browsers. *Lotus Notes*, Novell's *Groupwise*, and *Microsoft Exchange* are other examples of software supporting collaborative computing.

The need for collaborative computing is clear, as businesses and individuals must cope with more and more information and the cost of travel for face-to-face meetings continues to escalate. Employees spend too much time sorting through data in e-mail, faxes, mail, memos, reports, and voice messages. This problem is compounded by downsizing and restructuring in many companies, which translates to fewer people doing more work. Organizing the information and correctly forwarding it is also time-consuming. To stay competitive, businesses and organizations are turning to collaborative computing to share knowledge and resources and to move information efficiently.

A networked computer system provides the basis for a collaborative computing infrastructure. The software that makes up the collaborative computing platform (sometimes referred to as *groupware*) allows users to schedule meetings, coordinate calendars, send e-mail, work jointly on a document, or confer without physically being in the same geographic location.

3.8.2 Applications

From customer and account service to research and product development, collaborative computing can enhance many aspects of business. The most basic collaborative computing application, e-mail, has been around the longest. E-mail has replaced the written memo in many organizations, saving both time[8] (distributing the memo) and money (paper costs).

Collaborative computing can also simplify the process of filling out an expense report. Using an *intelligent form*, an employee need only enter expense amounts; the expense figures are then automatically calculated and the report is electronically submitted. After the form is automatically routed to the appropriate supervisor for review, it is electronically directed to the accounting department, which disburses payment. At any point in the process, the employee is able to track the report to determine its status. Similarly, purchase orders can be filled in and dispatched. This model permits fast and easy distribution, as well as convenient tracking.

Version control is another use of collaborative computing, in which the software makes it possible for more than one person to work on a document at the same time. The software keeps track of the latest version of the document, and updates all other copies as needed. Since hard copies of the document need not be sent back and forth, a large time savings is realized. If the system works properly, there is little chance that users will be out of synchronization while working on a document.

One of the most exciting applications of collaborative computing involves real-time interaction through *video teleconferencing* or simply *videoconferencing* (*VC*). Traditionally, business communication has involved the exchange of data and voice information; however, VC enables the real-time exchange of colorful video images and audio. While the potential uses for VC seem almost limitless, the most universal example involves the business meeting.

Businesses were the first to embrace VC technology, despite its high initial cost; they could justify their investment in terms of travel costs and time savings. Often, different groups of people in a single location need to

---

[8] Some would argue that e-mail only results in a shifting of time spent. E-mail speeds communication, but it can also increase the time spent in reading and writing messages.

communicate with other groups somewhere else. Thus, multiple VC equipment sets are necessary. Since the cost of good VC equipment is still fairly high, its use is not yet standard practice.

A less expensive technology for remote conferencing is *Desktop Video-conferencing (DTVC)*. DTVC uses regular personal computers and provides interaction between individuals situated at their own PCs, rather than between groups of people. Schools could benefit from DTVC by connecting teachers and home-bound students. A simple DTVC setup might consist of a PC connected to a miniature video camera through a video card. A microphone either could be connected through a sound card to the PC or might be part of the camera itself. More sophisticated DTVC systems contain the camera and microphone inside the monitor. Either way, a high-speed ISDN line, rather than a regular telephone or slower line, should be used to connect to the Internet, since transferring audio and video data requires much more *bandwidth* than transferring just data. (Bandwidth refers to the transmission capacity and is usually measured in bits per second.)

CUSeeMe is a free videoconferencing software tool that was developed at Cornell University in 1992. It is low bandwidth, so the quality is limited. CUSeeMe is available for both Macs and Windows systems. You can put together a very low-cost (under $200), nonproduction videoconferencing system using CUSeeMe. To receive, you only need a monitor that can display 16-level grayscale and an Internet connection; to send, you only need a (cheap) camera and a *digitizer*. You can find out more about CUSeeMe in the references.

It is worth emphasizing that a videoconferencing system may include any or all of the following aspects in order of increasing bandwidth requirements with varying technological costs.

1. Real-time talk or chat.

2. Whiteboard graphics.

3. Audio.

4. Black and white video.

5. Color video.

The costs of these technologies vary widely.

3.8.3 Impact

The major benefits of collaborative computing are convenience and time savings; these amount to money. Employees can examine, organize, and route data efficiently, while managers can have easy access to data and can find information in a timely manner. Electronically forwarding and accessing information saves time, since paper does not have to be physically distributed. The use of audio, graphics, and video plus text in a collaborative computing environment provides the means for clearer communications. This can result in fewer errors and misunderstandings. In addition, travel time and expenses can be significantly reduced by collaborative computing, since being in the same geographical proximity is no longer a prerequisite for the exchange of ideas.

3.8.4 Future Prospects

Collaborative computing may soon become a necessity for businesses that want to remain competitive. Unfortunately, a number of the commercial groupware products are quite expensive. However, some applications make use of Web technology and provide a cheaper alternative. For software that costs only a fraction of the well-known groupware products, some companies are utilizing a Web-based platform as their collaborative computing environment. In addition to being cheaper, the Web technology may be easier to use. Both Microsoft and Netscape include groupware in their version 4.x browser suites.

The infrastructure for a Web-based collaborative computing platform is an intranet. Access within an intranet is limited to employees and business contacts only, by a security measure known as a *firewall*. Web software developers are busy developing more sophisticated security measures and are quickly producing workable solutions. Intranet-to-intranet communication across the Internet is possible, if you use a technology that does not require dedicated bandwidth.

In response, groupware providers are trying not to compete directly with the Web technology. Instead, they are attempting to make their products compatible with the Web by allowing various browsers to access their databases. The ease and effectiveness with which groupware is able to meld with the Internet may determine its success. In the meantime, some companies are using a combination of Web technology and groupware. For example, an internal Web page may serve as a bulletin board or a means of displaying company manuals, while a product like Lotus Notes may be used for applications that require security.

Many businesses are already improving worker productivity through the use of collaborative computing. It seems clear that as developers overcome some of the current hurdles, collaborative computing will become even more prevalent.

EXERCISES 3.8 Collaborative Computing and the Internet

38. Describe three additional applications for collaborative computing beyond what was discussed in Section 3.8.2.

39. Research and report on the top three platforms for collaborative computing. Describe the major features of these platforms, including whether or not they are Web-based. Report on the prices of each. List your sources.

40.   Suppose that you were in charge of researching and selecting the groupware for your company, Antique T-Bird Car Parts, Inc., a mail order company that supplies genuine and replica car parts for antique Thunderbirds. Select a platform to use and describe how you arrived at this decision. Describe features of the platform that you think would enhance the productivity at your company.

41.   Would you want someone to see your calendar so they could schedule you into meetings? How would you feel about most of your office's communication taking place electronically?

The World Wide Web

4

4.1 Introduction

In this chapter, we examine the following:

OBJECTIVES

- The World Wide Web.

- Miscellaneous Web browser features, including a comparison of Netscape's and Microsoft's Web browsers.

- HTML writing styles.

- The outline, design, and management of a Web presentation.

- The *registration* of Web pages.

Much of what you will eventually learn about the Web and browsers will be self-taught. Our goal here is to help speed up this process by presenting key ideas. While the construction of quality Web pages is not easy, this chapter will give you enough background to put together a thoughtful Web presentation.

4.2 The Web Defined

The *World Wide Web* (WWW), or *Web* for short, is a software application that makes it easy and possible for nearly anyone to publish and browse hypertext documents on the Internet. You can think of the Web as encompassing all of the information available through Web browsers. The Web has developed a huge following because of its ease of use and visual appeal. It is called a "Web" because the interconnections between documents resemble a spider's web.

How does the Internet differ from the WWW? The Internet can be thought of as a very large group of networked computers. The information on the Web is transported over the Internet. The Web therefore uses the Internet. In fact, the Web is now the driving force behind the Internet. The Web generates the greatest traffic demands on the Internet, and new Internet technology is being developed to meet the needs and wishes of Web users.

Figure 4.1 shows the relationship between the Web, the Internet, and a number of other applications. Each application uses the Internet as a transport mechanism. The Web runs on the HTTP protocol. Browsers are *multiprotocol*, which means they can talk to many different kinds of resources that make up the Internet, and the number of protocols being included in browsers is steadily increasing. This illustrates an important distinction between the browser and the Web.

A central idea in the development of the Web was the *Uniform Resource Locator (URL)*, described in Section 2.3.2. A URL is a Web address that

FIGURE 4.1 The Relationship Between the Web, the Internet, and a Number of Common Applications.
(The particular geometric shapes are arbitrary.)

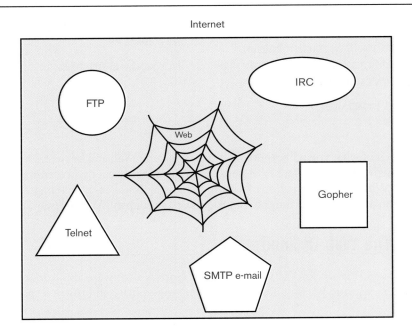

uniquely identifies a document on the Web. Such a document can be an image, an HTML file, a program, etc. Unique addresses make linking to anyone's Web documents possible.[1] The nonlinear nature of the Web is one of its main attractions. Since people all over the world are participating in and publishing on the Web, the entire system has evolved into something far greater than any individual contribution. The sheer amount of information available on the Web makes it the greatest collaborative nongovernmental effort in the history of humankind.

EXERCISES 4.2 The Web Defined

1. Assuming a 50 percent increase in the number of Web users per year, how many years will it take before you have more experience than all but 10 percent of Web users?

2. Write an HTML document summarizing your interest in the Web.

3. Many people find surfing the Web fun and entertaining. In your own words, try to explain why people find the Web so fascinating.

4.3 Miscellaneous Web Browser Details

In Section 2.2, we introduced browsers. Here we describe some other interesting features of these programs, including bookmarks, plug-ins, and helpers, and we compare Netscape's and Microsoft's browsers.

4.3.1 Personal Preferences

Most browsers have a number of options that you can set.

- **Cookies** You can ask to be notified before a cookie is written, and you can then decide whether or not to allow the cookie to be written.

- **Disk cache** You can set the size of your cache. The cache stores the HTML source code and images of Web presentations you have visited. Then, if you reload one of these pages, the browser can load the cached copy and the page will appear very quickly.

- **Fonts** You can select a font specification and also set a default font size.

---

[1] This is what a mathematician would call a *many-to-one mapping*. Each URL points unambiguously to a particular document, but multiple different URLs may resolve to that particular document, due to the use of aliases and shortcuts in the way names are assigned and used.

- **Helper applications**   You may configure *helper applications* to handle certain types of data that the browser is unable to process, for example, Word or PostScript files.

- **Home page location**   You can specify the initial page that gets loaded when the browser is first launched.

- **Images**   You may specify whether or not images are loaded. Options for color selection are also available.

- **JavaScript and Java**   You can enable or disable these types of programs from running within your browser.

- **Messages**   You may specify a default signature file or a default carbon copy address for outgoing messages. This can be set for regular e-mail or for posts to newsgroups.

4.3.2 Bookmarks

A convenient feature of browsers allows you to save the URL of any Web page you display. The saved URL is called a *bookmark* (or *favorite*). If you save the URL of a Web page, you have *bookmarked* the page. The bookmark is a pointer (URL) to a location on the Web, just as a physical bookmark inserted in a book is a pointer to a specific page. A collection of bookmarks is sometimes referred to as a *hotlist*, *list of favorites*, or simply a *bookmark list*. Since bookmark interfaces are usually mouse driven, selecting a bookmark from the list merely involves a point, select, and click process. Since URLs are generally cumbersome and difficult to type accurately, the mouse-driven interface is very convenient. For example, typing

```
rw.warnerbros.com/ns1_indx.html
```

to send a Warner Brothers greeting card is a lot more time-consuming than simply selecting this URL from a bookmark list.

Many new users begin bookmarking a wide range of URLs when they find new and interesting Web pages that they would like to be able to return to later. New users sometimes fear that they may not be able to relocate a cool page, so they tend to bookmark more Web pages than they will actually return to.

When you bookmark a Web page, the title of the page (as specified in the title tag), by default, is what goes into your bookmark list. Fortunately, browsers allow you to change this entry. Thus, for Web pages with nondescriptive titles, such as "My Cool Page," you can specify an entry that will remind you what the page was about. For example, "My Cool Page" might become "Luca Tomba's Home."

The bookmark pull-down menu, where your list of bookmarks can be viewed, usually appears in the browser's menu bar. As you accumulate more bookmarks, which are displayed vertically, the list may require more than one screen to display. As a result, you will not be able to view all your bookmarks at once. Many browsers let you sort[2] your bookmarks. Given the wide range of titles people use on their Web pages, sorting by title is typically not very helpful. An appropriate way to manage bookmarks so that you can locate the Web page you want with the least amount of effort is to create folders representing various categories in which you have saved multiple bookmarks (and then sort the bookmarks within these folders).

Bookmark folders are created using the browser's bookmark editor. The editor is usually mouse driven and user-friendly. Probably the most important feature of such editors allows you to create a new folder and specify its name. As you accumulate bookmarks, you can organize them into categories and then move the relevant bookmarks into their appropriate folders. The move is done by "dragging and dropping." Although this organization process is initially time-consuming, it can save you lots of time in the long run, if you use your bookmarks frequently. Also, some of the newer browsers allow you to file a bookmark as you save it. This is a very useful feature.

The two main, and very different, reasons for saving a bookmark are:

1. It is a page you expect to visit often.

2. It is a page you may never find again, so you bookmark it on the chance you will want to go back.

For illustration purposes, consider how a fictitious computer user, Björn, might manage his bookmarks. (You can apply a similar approach, or a different strategy that works well for you. The goal is to arrange things so that you can access your bookmark list easily.)

Initially, Björn had between 10 and 15 bookmarks; even without much organization, they were easy to use. As this number grew to 30 and then 40, it became apparent that it would be more efficient to group related URLs into separate folders. By the time Björn could actually spend the time to create the folders and organize his bookmarks, he was up to 80 bookmarks. He created the following folders:

Abba	Architecture	Badminton
Cycling	Food	Friends
Hiking	Hobbies	Jobs
Miscellaneous	Parallel Computers	Project
Swedes	Travel	Universities
Work	Yes	Zoo

For the most part, each folder contained four or five bookmarks. For example, under "Food," Björn had bookmarked Web pages about beverages,

---

[2] That is, place the bookmarks in alphabetical order according to their titles. Most browsers let you sort all bookmarks within the same folder.

dessert recipes, junk food, and Swedish dishes. Under "Friends," he had bookmarked the page of his old college roommate, Roxette, and the pages of two friends from graduate school. His "Miscellaneous" category took care of items that did not fit into any other folder. If enough related items ended up in Miscellaneous, he planned to create a new folder and migrate those bookmarks into it. Figure 4.2 illustrates his organization.

Many of Björn's bookmarks fell under the "Travel" umbrella. He had lots of URLs about the United States, a number of URLs about Sweden, several about Norway, and a whole group about vacation destinations. Within the Travel folder, Björn decided to add five new folders:

Miscellaneous	Norway
Sweden	United States
Vacation Possibilities	

Over time, Björn's bookmark collection has grown. He now has about 350 bookmarks spread over 30 folders, with each folder containing up to 10 sub-

FIGURE 4.2 The Arrangement of Björn's Bookmarks.

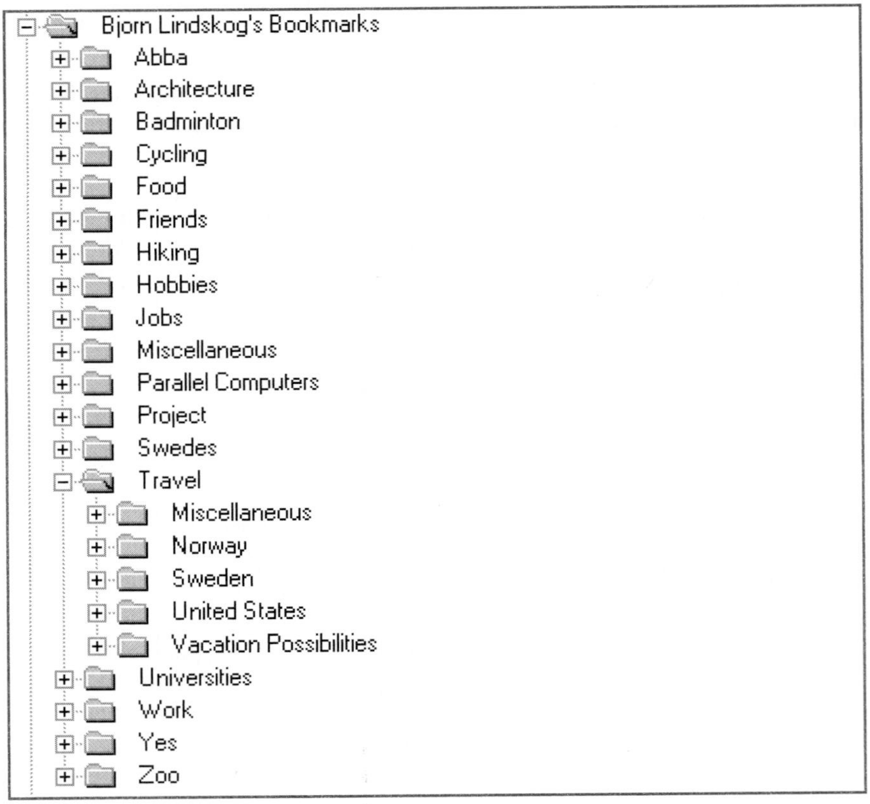

folders. When Björn views his bookmarks, all 30 of his main folders fit easily on one screen. He then picks a category and either selects a bookmark from there or enters a sub-folder. Most of his folders contain 10 sub-folders at most, and each folder rarely contains more than five bookmarks. This makes selection easy. Periodically, he finds it useful to delete or update bookmarks that have moved, and to migrate bookmarks from his various Miscellaneous folders, which appear in some sub-folders as well, into specific categories. When adding new bookmarks, Björn usually waits until he has a batch of five to ten before he places them into appropriate folders.

This example illustrates how organizing bookmarks periodically can cut down on maintenance time and improve efficiency. Although you may not want to be as methodical as our fictitious user, a half-hour here or there is time well spent in the long run.

Many browsers contain a feature called "What's New" that automatically tests your bookmarks and reports back which pages can no longer be found on the Internet. This is a great way to eliminate outdated URLs, or to prompt you to track down the URL of a page that has moved. Since URLs change so often, it is a good idea to use the "What's New" function. On average, every few months, when updating our bookmarks, we find that about 10 percent of our bookmarks have moved and another 5 percent are invalid. The entire checking process (for 300 bookmarks) takes anywhere from 10 to 30 minutes, depending on network traffic and the specific sites we currently have bookmarked.

4.3.3 Plug-ins and Helper Applications

Newly released browsers are equipped with even more features. New Web products and file formats are constantly being developed. When a browser is designed, it cannot handle every product and file format that currently exists or may exist in the future. Furthermore, browsers are not built to handle all existing data formats. Browsers are already huge programs, and bundling in more features makes them even larger. Yet, there are times when your browser is unable to handle a file that you would like it to be able to process, so you examine the possibility of extending the browser's capabilities.

For example, suppose you went to load a page and received a message such as:

```
no audio player installed
```

This message means that the Web page you loaded contains an audio component that your browser does not know how to play. You can fix the situation using one of two closely related mechanisms: *plug-ins* and *helper applications*.

Plug-ins and helper applications do the same thing; they extend the power of the browser. Plug-ins are more tightly integrated with the browser, so there is less work for you to do. All you have to do is put (drag) plug-ins into the browser's `plug-ins` folder so that it will find them and load them at start-up.

Be aware, however, that there is a memory cost to using plug-ins, especially memory-hungry ones such as *Macromedia's Shockwave*.

There are plug-ins that are capable of playing audio, showing movies, running animations, and working with calendars, among other things. Some plug-ins do no more than ease the running of a separate application, instead of being self-contained. The *Adobe Acrobat 3.0* plug-in is an example of such a program. Normally, plug-ins run in the browser window, but there are exceptions.

Netscape established and made public the programming interface for plug-ins, so anyone with the requisite programming skills and a C++ compiler can create one. More importantly, Microsoft now supports this plug-in standard, and Netscape plug-ins can normally be used with Internet Explorer and vice versa.

Helper applications, or *helpers*, are stand-alone programs that are used to process or display data that is not directly integrated into a Web page. Helpers do not display their information inside the browser's window; instead, they are launched in their own window. The browser activates the appropriate helper application when the browser encounters a file type that it does not know how to handle. Helper applications are usually configured in a list of preferences. You associate a file extension and MIME type with a particular software program. The helper application must be downloaded and placed in the appropriate directory.

Table 4.1 summarizes the features of plug-ins and helper applications. The choice of which ones to use will depend mainly on product functionality, product availability for your specific platform, and the memory requirements. You will need enough memory for your operating system plus your browser plus the plug-in or helper, all running at the same time.

4.3.4   Web Browsers Comparison: Netscape and Microsoft

The two most popular graphical Web browsers are *Navigator* from Netscape and *Internet Explorer* from Microsoft. The latest Netscape release includes

TABLE 4.1 A Comparison of the Features of Plug-ins and Helper Applications.

Plug-ins	Helper Applications
Closely tied to browser.	Program stands alone.
Displays inside browser window.	Displays in separate window.
Installation involves downloading the plug-in and running a procedure.	Configuring involves downloading the helper application and editing your preferences.
Broad selection available.	Broad selection available.
Launches quickly.	May launch slower than plug-in.

the browser, as well as many other features; the entire suite is referred to as *Netscape Communicator*.

Our goal in this section is briefly to compare and contrast these two applications. The best way to evaluate these two products is to try each one out for yourself.

The two browsers are actually quite similar, and they include many of the same features. Both browsers are large programs and are approximately the same size, depending on the exact versions. The number of HTML tags they support is large; however, you cannot assume that because your Web presentation looks great under Navigator, it will look equally impressive using Internet Explorer, or vice versa. Until now, neither company has truly embraced standards. Both companies have always incorporated some of their own features. For example, Netscape introduced tables and frames, whereas Microsoft supports marquees and a wide range of font faces. Such discrepancies have created challenges for Web designers who would like their pages to look equally good across all platforms.

The number of people using Internet Explorer is now nearly equal to the number of Navigator users. Netscape's market share is currently decreasing relative to Microsoft's, but not as fast as predicted. The trend is expected to continue. While both browsers are excellent products, the shift in market share is partially due to distribution capability. Netscape is usually downloaded off the Web or bought from a software vendor as a stand-alone product. However, Internet Explorer is packaged and distributed with a variety of other Microsoft products. The legality of such distribution is currently in question.

EXERCISES 4.3 Miscellaneous Web Browser Details

4. What are some personal preferences and corresponding values you like to set in your browser?

5. Some browsers have a bookmark feature called "Import." Describe what it does.

6. Describe your strategy for organizing bookmarks. Is it effective? How could you improve it?

7. What is the maximum number of bookmark folders your browser allows? How many bookmarks are allowed per folder?

8. Provide the names of three plug-ins and describe the purpose of each. Give the names of three helper applications and explain their use.

9. What platform seems to have the most plug-ins and helpers available? How many are there compared to other platforms? Does this give one platform an advantage over another?

10. Have you ever received a message like "plug-in not found," or something similar? What options did you have at that point?

11. Download a plug-in and test it. Describe the process.

12. Download a helper application and configure your browser to handle it. Test it and describe the process.

13. Describe two more differences between plug-ins and helpers.

14. Compare and contrast the Netscape and Microsoft browsers. Summarize your findings in a Web page. List your sources.

15. Report on the current status of the antitrust suits against Microsoft's Internet Explorer.

16. Have you used both Netscape's and Microsoft's browsers? If so, explain which you prefer and why.

17. (The purpose of this exercise is to have you conduct on-line research, alert you to the startling number of different versions of these products, make you aware of why the appearance of Web pages varies so greatly, and help you gain some insight into why it is so difficult to keep these products bug free.) How many different versions of Navigator are there? How many different versions of Internet Explorer are there?

18. Prepare a table that compares the sizes of Netscape's and Microsoft's browsers over a variety of different platforms. Does one product have an advantage over the other?

19. Prepare a table that compares at least five features of Navigator and Internet Explorer. Which product do you prefer?

20. As of today, how many people use Navigator as compared to Internet Explorer? Cite your source.

4.4 Web Writing Styles

The writing style required for a typical Web page is different than the writing style for the average printed page, because:

- Readers usually spend less time looking at a Web page than they do reading an off-line magazine or newspaper article.

- Web pages are typically very short, only one or two screens in length.

- Off-line material has greater longevity than on-line material.

- Web designers usually try to grab the reader's attention. If a presentation is not cleverly designed, a reader can easily move on.

- Web pages are hyperlinked documents, so readers typically do not go through them in sequential order.

- Published off-line material is generally written and edited by professionals, whereas Web pages may be published by anyone, sometimes with little or no writing and editing skills.

- Web pages are dynamic, and they often involve *multimedia*.

- With off-line material, the quality of the writing holds the reader's attention. The actual appearance and form of the writing are secondary. However, appearance and form are critical components of a Web page.

We can easily distinguish between publications that are designed primarily for on-line presentation (and secondarily for print) versus those designed primarily for print (and require conversion for on-line display). It is possible, however, to design for an intermediate or meta format that is then converted or filtered to both on-line and print versions.

Because of the inherent differences in writing styles for on-line and off-line material, some practice is needed to write specifically for the Web. We will consider a number of different genres that have become popular on the Web. The goal is to learn how to evaluate and become more aware of different Web writing styles, what works and why, and how on-line and off-line writing differ.

4.4.1   The Biography

Figure 4.3 depicts Paul Allen's biography from the Web. The page has a black background and is covered with information. If not properly laid out, a page as information packed as this one risks being too cluttered. The organization of this page, however, is excellent. The page has an image of Paul and numerous hyperlinks. The hyperlinks are displayed as a group of images, as well as clickable text. Each image incorporates a short, descriptive phrase explaining where the hyperlink leads. The page makes heavy use of multimedia (it provides an excellent demonstration of the *Shockwave* plug-in), and it offers the user a choice of pages, depending on the capacity of their connection.

What else can we learn about writing styles by looking at this page?

- This on-line biography contains up-to-the-minute information and is continuously being revised. Off-line biographies are quickly out of date, and it is impossible for them to remain as current as this presentation. This biography allows you to skip around, selecting what is most interesting to you. The material is written so that each hyperlink is more or less self-contained. Since the different subpages are not dependent, they can be read in any order. This is very different from a typical off-line biography, which is usually chronologically ordered.

- You can select any aspect of Paul Allen's background in which you are interested. The exploratory nature of the page greatly contributes to its effectiveness.

FIGURE 4.3 **A Sample Web Biography Page** (www.paulallen.com).

- The in-line images and animations help make the presentation fun to explore.
- The hyperlinks actually are the writing; most of the writing is contained in hyperlinks. In most other cases, you will find more text outside the

hyperlinks. The writing here is short and descriptive, but in most cases, even shorter hyperlinks are advisable.

The page is creatively done and contains a lot of information. The presentation meets its goal of providing readers with biographical information about Paul Allen.

4.4.2 The Business Exposition

Figure 4.4 depicts RubberChicken's business Web page. The page is brightly colored, and the text is very easy to read. Along the left margin is a narrow, colorful strip that includes a number of hyperlinks. This style is found on many Web pages. In this case, the hyperlinks are small, in-line images. The right-hand side of the page consists of a yellow background. The material in this portion of the page is formatted into a number of short paragraphs.

The descriptive title of the page is "FakeMail, pranks & gags: Rubber-Chicken.com." The header of the page consists of a graphic of the company name, RubberChicken.com, plus a quote from Victor Borge and an image of a rubber chicken. A menu that consists of a number of hyperlinks follows the header. Each hyperlink has a short, descriptive name. The menu is followed by information about the awards won by the company. At the bottom of the page, which is just over one screen long, is a footer that consists of an e-mail address for questions and also a copyright notice.

A careful examination of successful Web pages will help you gain new insights into effective on-line writing. Things to note in RubberChicken's page include:

- The images catch your attention. The menu is located in an easy-to-use spot, and very little must be read before a hyperlink can be selected. The images allow you to position your mouse quickly on information of interest to you. The narrow strip on the left has easy-to-use hyperlinks and interesting looking images.

- The page is compact, colorful, and easy to explore.

- The writing style is simple and direct. In our example, a number of facts about the awards are presented, but there is very little additional writing. Short, descriptive, and consistent-looking hyperlinks are included in the menu. The majority of people are skimming and scanning on most Web pages, not reading every word.

- The page has a descriptive title, an effective header, a short footer, and a centrally placed well-designed menu. Such ingredients create a high quality Web page. The page is just over one screen long and therefore does not require much scrolling.

FIGURE 4.4 A Sample Web Business Page (www.rubberchicken.com).

As evidenced by the large number of awards[3] this presentation has won, the writing style it uses is very effective. The presentation designers met their goal of having a lot of visitors enjoy their pages.

---

[3] Awards, while seemingly impressive, should be examined as to their criteria and the people applying these criteria.

4.4.3   The Guide

Figure 4.5 depicts a page called the Ultimate Band List guide. The page starts with a descriptive title, "Welcome to the Ultimate Band List." The header is an image map that appears at the very top of the page. The image map provides hyperlinks for easy navigation. The descriptions on the hyperlinks are all short, informative, and consistent. A number of brief paragraphs then describe activities in which you can participate. These are offset from each other by horizontal image bars. The footer consists of a number of clickable graphics, one of which is set to an e-mail address where you can pose questions about the information.

By incorporating the best features you see into your Web pages, you will be able to produce a more effective and informed presentation. For our example:

- The page provides convenient information about bands. The search mechanism in the header allows you to look for the band of interest immediately. In addition, the page has a huge "database" of bands, so there is a good chance you will turn up the group in which you are interested. By being easy to use and comprehensive, the page delivers on the promise in its name.

- The graphics are well done, the page is easy to read, and the hyperlinks are easy to follow. You can make the guide work for you without effort. The page is even easier to use than a dictionary.

- The image map at the top of the screen and the search mechanism allow you to locate the information you are after quickly. Other hyperlinks are effectively blended into the writing. At the bottom of most screens, there is a hyperlink to return you to the presentation's main page.

- The color scheme is appealing, and the text is very easy to read on the white background. The choice of colors and the preparation of graphics were carefully done. Notice the alternating-color horizontal bars that separate the different sections.

The presentation is coherent, informative, and easy to use. It creates an uncluttered, user-friendly environment.

4.4.4   The Tutorial

Figure 4.6 depicts the Cookie Central tutorial page that explains everything you ever wanted to know about cookies. The Cookie Central presentation is visually appealing; its color scheme works well. The title of the presentation is a simple and descriptive "Cookie Central." The header consists of a graphic that repeats the title of the presentation. This is useful in that if someone makes a hardcopy of the page, the title will be included.

The left-hand side of the page consists of a separate-colored, narrow strip. The strip contains a *select menu*, from which you can choose a topic to follow. The right-hand side of the page is formatted with an HTML table. The

FIGURE 4.5 A Sample Web Guide (ubl.com).

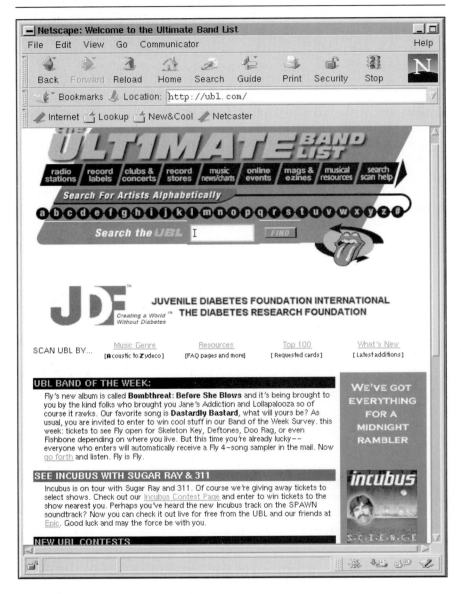

The Ultimate Band List Internet site is owned and operated by ARTISTdirect, LLC.

first column of the table contains clever phrases, with associated small icons. A hyperlink is included as the "heading" of the corresponding phrase. The right-hand column of the table consists of a list of hyperlinks. The hyperlinks are short, descriptive, clickable text.

FIGURE 4.6 A Sample Web Tutorial (`www.cookiecentral.com`).

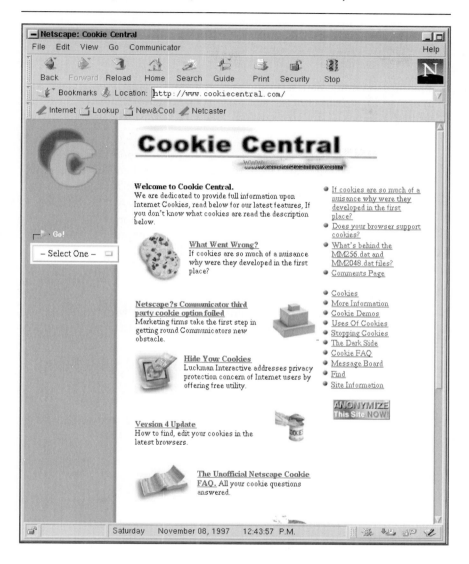

The page is about two screens long, and it concludes with a footer that has a hyperlinked copyright notice. This hyperlink takes you to information about the presentation and provides `mailto` links for sending in comments or questions about the material.

- The presentation, although packed with information, is not intimidating. The text phrases are easy to read. The select menu allows you to decide what topic to pursue next. All of these features combine to create a very user-friendly presentation.

- The writing is short, descriptive, and light but informative. The graphics are visually appealing. It is easy for you to explore the presentation in a nonlinear fashion.

- This presentation is far less intimidating than most computer manuals. The information is clear, concise, and easy to understand. Information in the tutorial is much easier to find than it would be in most other forms of computer documentation.

- The graphics spruce up the appearance of the page. In addition, the images suggest what pages they lead to, as the in-line images themselves are hyperlinks. The graphics are lighthearted, yet they lead to serious information.

The goals of this tutorial are met because the presentation is fun, and learning from it is easy. Contrast this with your typical computer manual. The consistency of the presentation adds to its overall design.

4.4.5 Writing Genres Summary

In the last few sections, you have explored four different writing genres: the biography, the business exposition, the guide, and the tutorial. Some key points that made these presentations effective are:

- A theme and clear goals for the presentation.
- Ability to permit exploration, while providing sufficient navigation so that you can easily avoid getting lost.
- Good choice of colors, with the text easy to read.
- Consistent and careful page design, with hyperlinks normally situated near the top of the page.
- User-friendly navigation.
- Short, descriptive writing.

 Other genres will also contain many of these key ingredients. It is a good idea to include most of these items in your own presentations.

EXERCISES 4.4 Web Writing Styles

21. List three important similarities between off-line and on-line writing styles.

22. Describe two other writing genres that are popular on the Web. Identify URLs to presentations that exemplify those genres.

23. List two URLs for each of the following writing categories:

    (a) Biography.

    (b) Business.

    (c) Guide.

    (d) Tutorial.

24. Can you locate any on-line writing guides for Web page development? List their URLs. Report two interesting facts you learned.

25. Locate a Web presentation that interests you and critically evaluate how well it accomplishes its goals, by posing and then answering a number of questions about it. Carefully assess how well the Web author attended to details.

4.5 Web Presentation Outline, Design, and Management

A *Web presentation* is a collection of associated, hyperlinked Web pages that usually have some underlying theme. Scan a number of presentations on the Web and you will see a radical difference in the quality of those presentations. Most users have an innate ability to distinguish between an effective presentation and one that is poorly done. Why is it so easy to get an accurate first impression of a Web presentation? Because Web presentations are so visually oriented, you can often get an idea of the author's capabilities just by glancing at a few pages.

Some of the key elements in an effective Web presentation are:

- **Details** A presentation that includes well-thought-out touches can make a positive impression on the reader. Good choices of background color, headers, footers, and appropriate font size are all important. Typos, poorly aligned images, and a clashing color scheme create a negative image. Carefully prepared pages help to enhance the credibility and readability of the presentation.

- **Coding** A competent HTML programmer uses the appropriate elements of style, not to show off, but to contribute to the overall quality of the presentation. Overly simplistic pages could indicate a lack of knowledge by the designer, and cluttered pages indicate poor design.

- **Features** A few "bells and whistles" can improve the feel of a set of pages. For example, *splash screens*, *applets*, or some *JavaScript* can spruce things up. However, it is better to include too few fancy items than to go overboard. Remember that each "cool" feature takes extra time to download and display.

- **Graphics** A modest use of in-line images is probably one of the most significant ways to enhance your design. Many free images can be found on the Web. It is important to integrate the images effectively into the presentation. Randomly positioning images usually does not enhance the design.

- **Layout**   If the pages are visually appealing and if they provide convenient navigation, they are likely to be visited by more users. A poor layout will discourage people from spending too much time on the presentation.

- **Writing quality**   Good writing and an interesting style are also necessary for a solid presentation.

- **Load time**   Since many users have 28.8 modems, care should be taken not to include too many graphics or any large (in bytes) graphics. Also, users should be given the option of downloading large images separately. If a user gets frustrated during the loading process, they are likely either to stop loading the page altogether or be more critical of the page when it finally arrives. Single pages should be limited to about one or two screens in length.

- **Hyperlinks**   Navigation is a key element to any good Web presentation. If the reader can move around the pages easily, they will be more impressed. If they can easily get lost and are not able to find the desired information, they may become frustrated. However, it is also possible to include too much navigation. For example, some presentations include an icon of a house, an image of an index finger, and the phrase "Return Home" to indicate how to go back to the top-most page. Any one of these items would suffice.

The sample writing genres covered in the last section all receive high marks for each of these criteria.

4.5.1   Goal Setting

The important steps in producing a high-caliber presentation are goal setting, outlining, navigating, designing, coding, and revising.

The object of goal setting is to decide on specific parameters for your Web presentation, to aid you in focusing your project. Most people find it helpful to write down their goals. The written record allows you to go back and see how successful you were at accomplishing your goals. A number of issues for you to consider before creating your Web presentation are discussed here. Each project will merit its own additional questions.

- *Audience*—Ask yourself who the intended audience is. Are you writing for children, single moms, business executives, retired persons, or a general audience? Are your intended visitors scattered throughout the world, or are they all on the same LAN? Is your goal to have a certain number of hits per month? Knowing who you are targeting will greatly influence your design.

- *Date*—What is your timeframe? Many things published on the Web need to be displayed in a timely manner. Is your goal to finish your presentation by the end of a school semester? Time constraints will impact the depth and extensiveness of your project.

- *Graphics*—How many graphics do you intend to include? Do you want to impress your audience? Are you going for a conservative approach? Remember, graphics take a long time to download. If your audience is local, you can use more graphics than if you are targeting a global group. Also, custom graphics take a long time to design, so your schedule will impact the number of graphics you can create.

- *Length*—How much material is going to be included in your presentation? Is your goal to produce a comprehensive presentation? In contrast, do you want just to touch on a few key ideas and provide references to other, more in-depth presentations? Your time constraints will also impact how much material you can develop.

- *Maintenance*—Is your goal to design a presentation that does not require any maintenance, or is the nature of the material such that items will need to be modified from time to time? How thoroughly you comment your code may depend on who will be maintaining the presentation. Do you need to build in flexibility to accommodate future changes?

- *Money*—Are you trying to earn money from your presentation? Is your goal to impress people to obtain other jobs? Are you building an on-line gallery? Such factors will impact your design.

- *Objective*—What is your main objective? Are you designing your presentation for personal satisfaction, for a class, for a friend, or for a business? Write a couple of sentences that summarize the most important points about your presentation.

- *Research*—Does your project require a lot of research? Can the necessary research be accomplished on-line? How many sources do you want to include? Are you going to incorporate a history of what other people have done? Alternatively, does your presentation not need any in-depth research? Again, your timeframe will influence the level of research you can conduct.

- *Writing*—Is your goal to have accurate and error-free writing? Do you want high-quality prose? Do you want visitors to read everything you write? Are you trying to teach your readers about a certain subject? Set goals for what you would like your writing to accomplish.

Before designing a Web presentation, run through this checklist; it will help you get started and focus more clearly. As you develop more and more presentations, modify this list to suit your needs better.

4.5.2 Outlining

Once you have set goals for your Web presentation, it is a good idea to produce an outline. The subject matter, combined with your goals, will dictate the way the material is most naturally partitioned. Most students work best if they design their outline in HTML, rather just handwriting it on paper. Creativity seems to be enhanced when working on-line. Actually seeing the outline on-line seems to help students decide what is missing, what needs to be deleted, and what needs to be moved.

There are different schools of thought on the Web page development process. Ours is to produce the HTML code concurrently with the actual writing. Others believe in a two-step process:

1. Develop the writing for the presentation.

2. Then code it in HTML.

These two steps can be completely independent, but our experience shows that writing and coding together (not independently) leads to an incremental development process, and typically a better design. When a writer experiences a "mental block," they can focus on coding. Seeing the presentation take shape on-line can also stimulate more thoughts about what to write. Similarly, writing something and seeing it on-line can generate ideas on layout. Also, observing the incremental progress is rewarding; it shows that the effort is paying off. This can be much more satisfying than producing a few paragraphs on a sheet of paper. Staring at a piece of paper with a lot of cross-outs does not give you the same sense of achievement and satisfaction that a couple of paragraphs coded in HTML and viewed on-screen does. For these reasons, we recommend coding and writing simultaneously.

The best way to describe how an outline is constructed is to make one. For our sample outline, we used the process that describes "how to clean a typical dorm room." The goal is to provide a set of instructions about cleaning. Suppose, for this example, that the dorm room consists of a bedroom, a bathroom with shower, and a small kitchen. Figure 4.7 illustrates the features of the room.

We begin by listing the main items and creating a title for the page.

```
<HTML>

<HEAD>
<TITLE>How to Clean a Dorm Room</TITLE>
</HEAD>

<BODY>
<H1>How to Clean a Dorm Room</H1>

<H3>Bathroom</H3>
<H3>Bedroom</H3>
```

FIGURE 4.7 The Dorm Room Used to Illustrate the Outlining Process.

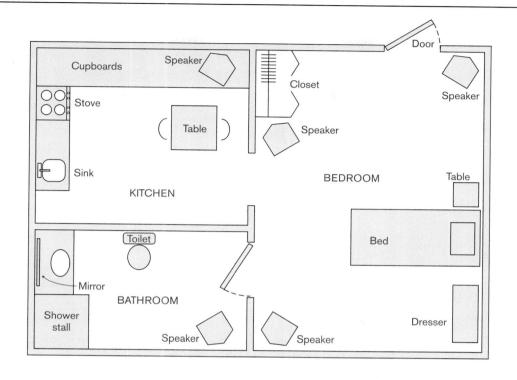

```
<H3>Kitchen</H3>

</BODY>

</HTML>
```

Notice in the code that we simply listed the names of the different rooms, and we chose to put them in alphabetical order. We also chose to capitalize the words in the title, using the rules of grammar. Observe that we listed the title of the document twice, once in the title tag and again as part of the body of the document. In the latter case, we used an `<H1>` level heading. The primary reason for repeating the title is so that it will be included on a hardcopy. We chose a modest size, `<H3>`, for our section headings. Figure 4.8 shows how the HTML code is rendered at this point.

Thus far, we have created a simple HTML document and a subheading for each of the three "rooms." Next, we expand the outline by inserting the subtasks into each room heading. This really amounts just to listing the items in the room. For example, for the bathroom, we need to include the floor,

FIGURE 4.8 The First Cut at an Outline That Describes How to Clean a Dorm Room.

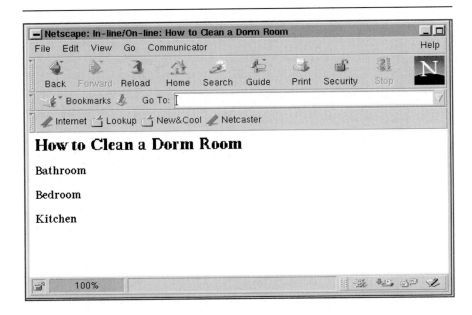

mirror, sink, shower stall, and toilet. For the bedroom, we have to deal with the bed, closet, dresser, floor, and table. For the kitchen, we list the countertops, cupboards, dishes, floor, sink, stovetop, and table. When we add these items into our outline, we will include some descriptive language to indicate exactly what needs to be done.

Here is a second cut at the outline. Observe how we have written the code to suggest further the structure of the outline.

```
<HTML>

<HEAD>
<TITLE>How to Clean a Dorm Room</TITLE>
</HEAD>

<BODY>
<H1>How to Clean a Dorm Room</H1>

<H3>Bathroom</H3>

<H5>Mop Floor</H5>
<H5>Windex Mirror</H5>
<H5>Clean Sink</H5>
```

```
<H5>Scrub Shower Stall</H5>
<H5>Scour Toilet</H5>

<H3>Bedroom</H3>

<H5>Change Sheets and Make Bed</H5>
<H5>Hang up Clothes in Closet</H5>
<H5>Straighten up and Close Dresser Drawers</H5>
<H5>Vacuum Floor</H5>

<H3>Kitchen</H3>

<H5>Wipe Down Countertops</H5>
<H5>Wipe Down Cupboards</H5>
<H5>Wash, Dry, and Put Away Dishes</H5>
<H5>Mop Floor</H5>
<H5>Clean Sink</H5>
<H5>Wipe Down Stovetop</H5>
<H5>Wipe Down Table</H5>

</BODY>

</HTML>
```

Note that for the subtopics related to each room, we have used `<H5>` headings. Figure 4.9 depicts our second pass at the outline.

At this point, the outline is pretty solid and will suffice for our purposes. Of course, items like "put away cleaning supplies, deodorize the rooms, and take out the garbage" could also be added. The key to becoming good at outline design is practice. The outlining process is partially dependent on the navigation process explored below.

We have skipped over one detail, debugging. Our code "came out correctly." While you are coding, however, it is a good idea to enter only a small amount of new code and then view it using your browser, to ensure that everything works correctly. Check to make sure you ended all tags appropriately, correctly spelled all tag names, used valid tags, and so on. The debugging process can be simplified by this approach, since errors are much more localized if only a few changes are implemented at a time. In Netscape 3.x or higher, if you have a syntax error and use the View Source option, the browser will flag the error with a special color and blink.

When you create your first Web pages, you might consider them to be expendable. Their purpose is to practice HTML and prototype. We should

FIGURE 4.9 The Second Cut at an Outline That Describes How to Clean a Dorm Room.

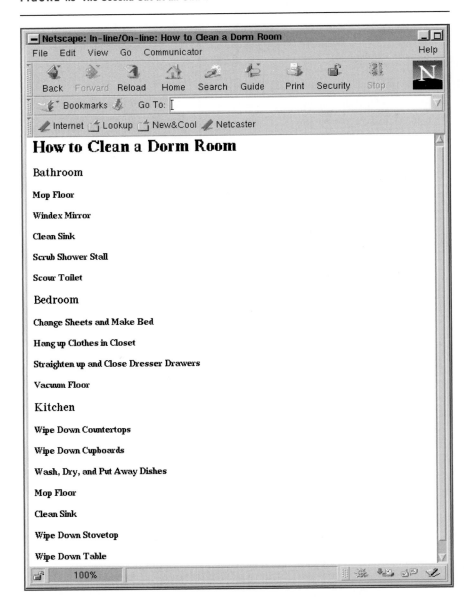

also note that you do not need a Web server to view your pages during development. They can be viewed from your own computer using your browser, and can then be installed when they are ready.

4.5.3   Navigating

The *navigational tools* you provide are hyperlinks that allow your readers to move more smoothly through your pages. Readers should be able to jump over material that is uninteresting to them or move directly to any of the main sections. It should be convenient for them to return to either the main page or an index, which can easily be crafted from your outline.

The organization of a Web presentation will dictate the nature of its accompanying navigational aids. Several general organizational arrangements have become popular.

- *Circular*—The circular arrangement of Web pages supports forward and backward movement through the pages. This format is especially good for describing step-by-step procedures or instructions or for dividing up text that should be read sequentially. Figure 4.10 illustrates a four-page circular organization.

  In Figure 4.10, the four pages are labeled A, B, C, and D, where A is the starting page and D is the ending page of the presentation. The hyperlinks are shown as arrows pointing to other pages. From the entry page A, a user can move to B, C, and then D. From D, they can return to the beginning

FIGURE 4.10 A Four-Page Circular Web Presentation.

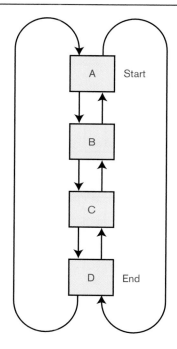

by using the hyperlink to A. Notice that the hyperlinks from A to D can be followed in reverse order, to go from D to A.

Conceptually, the pure circular organization is very clean. Users have little trouble following the hyperlinks, and they always know where they are in the presentation. One drawback to this approach is that there is no way to jump to the middle of the presentation. However, since circular organizational structures are usually reserved for sequential presentations, this is not a serious hindrance. Furthermore, at the risk of introducing a bit more complexity to the navigation, hyperlinks to the middle of the presentation could be added when many pages are involved.

- *Exploratory*—The *exploratory arrangement* allows jumping from nearly every page to nearly every other page, in an order completely determined by the user. This form is similar in structure to a spider web, and it works best for describing things like geographical areas or maps. Figure 4.11 depicts a nine-page exploratory arrangement.

    The first thing you notice about the exploratory arrangement shown in Figure 4.11 is that there are a lot of hyperlinks. This gives the reader the freedom to move about as desired, but it also provides many opportunities for becoming lost. The web-like structure has no real beginning or ending,

FIGURE 4.11 A Nine-Page Exploratory Web Presentation.

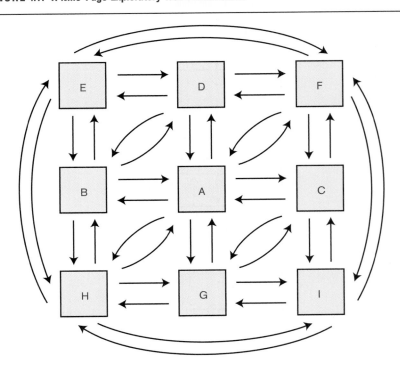

so it may be difficult to determine where you are in the presentation. Yet, for an on-line tourist map of a city, this type of navigation works best. However, for presentations of a more sequential nature, the exploratory arrangement may confuse the reader.

Another issue in the exploratory model is the effective display of the hyperlinks. For example, notice that page A has six outgoing hyperlinks. For a map, such names as N, NE, E, S, SW, and W might be effective for these links. In any case, the hyperlinks should be kept short and descriptive; otherwise, they may overwhelm the reader. *Image maps* work well in the exploratory type of organization.

- *Hierarchical*—The *hierarchical arrangement* permits a more limited number of hyperlinks from the introductory page, and each succeeding page leads to additional hyperlinks. This arrangement yields a tree-like or directory structure. The *fanout*, defined as the maximum number of hyperlinks available from any page, should be kept small. In general, fanout should be no more than ten hyperlinks. The *depth* of the presentation, defined as the number of levels of the tree, should be restricted to five at most.

  The hierarchical model is well suited for describing companies, institutions, and organizations, that is, entities that have an inherent hierarchy. For a typical presentation, values of three for fanout and three for depth should suffice. Figure 4.12 illustrates a 13-page hierarchical layout.

  The arrangement shown in Figure 4.12 is very regular; that is, from each page on the second level, there are three sub-pages; each page on the third

FIGURE 4.12 A Thirteen-Page Hierarchical Web Presentation.

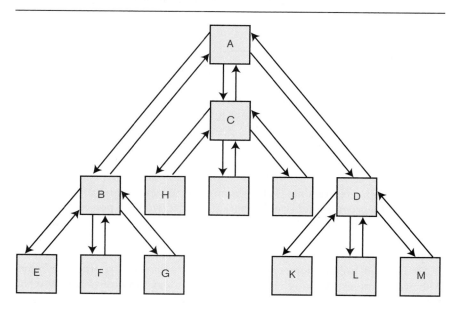

level has just one upward hyperlink; and so on. Obviously, such a regular structure makes it easy for the user to anticipate the navigational options that will be available on each page. Many organizational structures are not quite so regular. However, it is easy to modify a regular hierarchical arrangement to meet the organizational needs of most presentations, without giving up navigational ease.

As you navigate different Web presentations, pay attention to the type of organization and navigation that you feel works best. Try to incorporate these designs into your Web presentations. Before embarking on the development of your own presentation, plan a global navigational scheme.

In addition to the structure the user sees in your pages, you must decide on the physical organization of your files into directories and subdirectories. Suppose you were developing a page for your organization and you wanted to include a list of members. You might put all the biographies in one directory. If you had pictures of members, you might put those in yet another directory. The rule is to divide the material into as many different directories as makes sense.

Another practical issue is the way you name your files. Aim for consistency, and try to avoid simple but insidious coding errors. For example, on most modern operating systems, case is significant in file names. We recommend that you always use either upper or lowercase, but not a mixture, with some files uppercase and others lowercase. People sometimes capitalize the first letter of directory names and treat regular files as all lowercase. Even though the file system may allow blanks in file names, do not use blanks in Web file names—they get rendered as "%20" in URLs. If you use hyphens or underscores, use one or the other, but not both.

4.5.4 Designing and Coding

To write an effective Web presentation, set your goals and develop an outline and a global navigational design. Producing an HTML document from a design involves coding, as well as writing. Therefore, it is necessary to be competent with HTML before attempting to develop a serious Web presentation. You need to know what HTML provides in order to take full advantage of it.

The HTML tags presented in this book will provide you with the capability to produce a quality Web presentation. Some might argue that a decent writer could develop a reasonable Web presentation; others might suggest that a good HTML programmer could generate an effective Web presentation. We believe that neither of these skills is sufficient by itself. The writing and coding are both essential, and they should take place in parallel. This is true whether the presentation is being developed by an individual or a group.

Many good Webmasters acquired some of their skills by studying other Web presentations. Although beneficial, this strategy can only take you so far.

Our goal is to provide you with the means to design, develop, and code your own presentation. The design and coding process that we have found works best is as follows:

1. *Navigating*—Use your list of goals and your outline to design a navigational strategy based on one of those presented in Section 4.5.3, or modify one of those to suit your needs better, or formulate a new navigational scheme. Once the strategy is designed, add the HTML code. Set up the appropriate number of pages and the hyperlinks between them. Migrate the relevant parts of the outline into each page, but keep a copy of the original outline. Use section headings from the outline as titles for the pages, or assign new titles. Design, code, and insert *headers and footers* into each page, as discussed in Section 7.3. You should now have a shell for the entire presentation. You should also be able to navigate through the presentation and get a good feel for what the final result will look and feel like.

2. *Coding and writing*—With the outlining and coding you have done so far, you have created places for the writing. Our experience shows that writing flows more readily when writers have a good sense of what and how much they need to write. The shell you have developed should provide a reasonable direction for the text.

   If you hit a writing block, continue coding. When you are not sure how to code something, do more writing. Periodically view what you have created. This will help with the debugging.

3. *Revising*—After you have completed the first draft of the presentation, if you let it sit for awhile, you will probably come back to the project with a new perspective. It is then time to begin the revising process.

4.5.5   Revising

Having completed the first draft of your Web presentation, you have decided on a topic, set goals, outlined the project, developed the navigation, done some designing, and then written, coded, and debugged the material. Fine-tuning is the crucial step for converting an average presentation into an excellent one. The goal at this point is to move forward with the remaining work of perfecting your presentation.

Prototyping, which has played an important role in computer science, involves designing a system to work out the kinks and learn how the system should really be built. Sometimes the prototype is discarded and a completely new system is built. The main idea is to keep revising until you are satisfied with the final product.

If you decide to solicit feedback from others, explain your goals and ask for critical evaluations of your presentation. Other viewpoints can provide constructive ideas that can help you to improve your presentation. As you become more accomplished at Web design, the revision process will require fewer iterations.

Some of the issues to consider when revising a Web presentation are as follows:

- *Check navigation*—Test all hyperlinks. Make sure the overall navigation is appropriate and easy to use. Critically evaluate the descriptions of the hyperlinks themselves, using the criteria established in Section 2.7.3.

- *Comment code*—Thoroughly comment your code so that on a later revision, you can recall what you did and why.

- *Final evaluation*—Did you meet your goals? Did you complete all of the tasks from the outline? If not, address the issues that are still outstanding.

- *Graphics*—Are the graphics you used necessary, are there enough graphics, and do they enhance the presentation? Do the graphics load fast enough? Are the graphics properly positioned? Would they look better elsewhere?

- *Headers and footers*—Make sure all headers and footers are consistent.

- *Reread*—Critically evaluate your writing. Check that everything makes sense. Pay attention to the little details, such as consistent spacing after punctuation, consistent capitalization, and so on.

- *Spell check*—At a minimum, run the pages through a spell-checking program.

The better the overall design, the less time will be necessary to manage the completed presentation. Management of a well-designed presentation should only involve updating. As you gain more experience, you may want to modify this revision checklist to suit your personal needs better. Learn your own strengths and weaknesses, so that you can better focus your attention.

EXERCISES 4.5 Web Presentation Outline, Design, and Management

26.   Construct a humorous Web page containing an exaggerated use of images and phrases indicating how to return home.

27.   Create a Web page outlining the rules of your favorite board game. Among other things, use various heading tags.

28.   Outline the process of obtaining a college degree in your major. Put your description into a Web page that uses a number of heading tags for various section levels.

29.   In HTML, outline the process of obtaining a credit card.

30. Produce an HTML outline describing the process of eating in a dining hall.

31. Describe two other possible organizational models for a Web presentation. (The text describes circular, exploratory, and hierarchical.) When might they prove useful?

32. Draw a circular organizational scheme with three pages. In a circular organization consisting of n pages, what is the maximum number of hyperlinks you would need to follow to reach any other single page?

33. Draw a five-page exploratory organization of pages in which it is possible to go directly from each page to every other page. How many hyperlinks are involved? Can you devise a formula, as a function of the number of pages, for the number of hyperlinks involved in such a "complete" scheme?

34. How many pages would a fanout four and depth five hierarchical scheme be able to accommodate? How many hyperlinks are involved?

35. List three management tasks that an ongoing Web presentation might require. Elaborate on each one.

4.6   Registering Web Pages

Registering a Web page means having the page *indexed* by a search engine or having other prominent pages display hyperlinks to the page. The goal is to have the page viewed by more users. If a search engine indexes a page, then the search engine can return the page to users' queries. The more search engines that know about the page, the more likely the page will be accessed. Similarly, having several other pages add hyperlinks to a page increases the likelihood that the page will be visited.

How do you register your Web page? Many search engines allow you to fill out and submit an on-line form telling the search engine about your page. Search engine designers want their programs to know about more Web pages than any other search engine. Program performance is judged by the speed of the search and the number of *hits* a search engine finds. A hit means the search engine found a page that matched a query. By registering your Web page, you are helping a search engine expand its knowledge base. Most users register their pages with many different search engines.

Search engines may index your page automatically, since they have techniques for "going out and seeing" what is currently available on the Web. This begs the question of whether search engine registration is necessary, in the sense of proactively seeking registration, rather than waiting to be found eventually. Adequate publicity may occur in a group newsletter or a posting on an e-mail list.

There are some businesses on the Web whose pages get visited frequently. Usually, a business will add a hyperlink to a specific page, for a fee. This service is most commonly used by people with commercial pages. Several nonbusiness-related pages also receive a tremendous number of hits, and under the appropriate circumstances, you can ask them to include a hyperlink to your page. For example, they might find your page interesting and complementary to their pages. There are also noncommercial, nonsearch engine sites that are interested in having hyperlinks to pages on a specific topic. For example, someone may have the definitive "Theta Gamma Pi" page and be interested in a hyperlink to your Greek chapter.

There are also directory services that you can notify about your page. Such services are generally subject-specific or may pertain only to a certain group. For example, there is a Web page where Appalachian Trail thru-hikers can register their Web pages. Other groups have indexes where a member can add a hyperlink to their page. Once listed in the directory, a page is more likely to be visited.

Another consideration to registration and publicity is *traffic*. The person responsible for the server on which your pages reside may not be prepared to handle an enormous bulge in traffic. Consult with your server manager if you have a Web page that you feel could become very popular.

The challenge is knowing where to register your page to obtain a larger audience. If your goal is to maximize the number of hits to your pages, register the pages in as many places as possible.

E X E R C I S E S 4.6 Registering Web Pages

36. Does your school or organization have a listing of Web pages where you can register yours? If so, describe the registration procedure. If you would like, register your page with them.

37. Provide the URLs of three businesses that will add a hyperlink to your Web page from their presentation. What are the costs of doing this, and how many hits do the businesses claim to receive per week?

38. (This exercise illustrates the unlikelihood of someone randomly coming across your page.) Suppose Tammy Howard is able to visit one Web page every 15 seconds, and there are a total of 1,000,000,000 Web pages. Using a random search strategy, how long will it take before Tammy has at least a 50 percent chance of visiting your page?

Search Topics

5

5.1 Introduction

With the advent of the World Wide Web came the widespread availability of
on-line information. It is no longer necessary to travel to the library to find
the answer to a question or engage in research on a specialized topic. Much
of what you might want to know is available through the Web. Since anyone
can publish on the Web, the range of topics that can be found is nearly all-
encompassing. However, while a lot of information is available on-line, not all
of it is completely accurate. In Section 3.6.1, you learned that it is necessary
to evaluate information on the Web very critically.

In all likelihood, the answers to your questions are somewhere on the Web,
but how do you locate them? In the early days of the Web, unless you knew
exactly where to look, you had trouble finding what you wanted. Unlike a
library, the pages on the Web are not as neatly organized as books on shelves,
nor are Web pages completely cataloged in one central location. Even knowing
where to look for information (that is, knowing a URL) is not a guarantee that
you will find it, since Web page addresses are constantly changing. Usually, a
forwarding address is provided for a page that has moved, but it may only be
available for a short time.

The rapid growth of the Web, as well as its huge size, has ruled out trying
to keep track manually of "what's what" and "what's where." As people were
spending their time trying to find things on the Web, rather than actually
reading the material they were after, the first *directories* and *search engines* were
being developed. These tools allow you to find information more quickly and
easily. You have probably already been using these tools, but perhaps not as

effectively as possible. In this chapter, you will learn how to use them more efficiently, as well as how they work. Our goals here are:

OBJECTIVES

- To describe *directories*, *search engines*, *metasearch engines*, and *white pages* and present examples of each.

- To discuss search fundamentals, such as *query* types.

- To describe search strategies, using a number of sample searches.

- To examine how search engines work.

5.2 Directories, Search Engines, and Metasearch Engines

Two basic approaches have evolved in response to the need to organize and locate information on the Web: directories and search engines.

5.2.1 Directories

The first method of finding and organizing Web information is the directory approach. A *Web directory* or *Web guide* is a hierarchical representation of hyperlinks. The top level of the directory typically provides a wide range of very general topics, such as arts, automobiles, education, entertainment, news, science, sports, and so on. Each of these topics is a hyperlink that leads to more specialized subtopics. They in turn have a number of subtopics, and so on until you reach a specific Web page.

In addition to being very easy to use, another benefit of a directory structure is you need not know exactly what you are looking for in order to find something worthwhile. You select (click on) the category for the topic in which you are interested. You continue to move down through the hierarchy, selecting subcategories and narrowing the search at each level, until you are presented with a list of hyperlinks that pertain to your topic.

As you begin to zero in on your topic, you may find other interesting items of which you were previously unaware. On the other hand, you may reach the bottom of the directory without finding the information you were after. In such cases, you may need to backtrack, going up several levels and then proceeding down again. Of course, it is possible that the directory you are searching does not contain the information you want, in which case you may decide to try either a different directory or a search engine.

When traversing a directory downward, you are moving toward more specific topics. When going upward, you are heading back to more general topics. Directories are useful if you want to explore a topic and its related areas, or if you want to research a subject, but not at a very detailed level.

If you are interested in a very specific topic, you may want to start off by using a search engine or a metasearch engine. Arriving at a very specific topic in a directory structure involves traversing between five and ten hyperlink levels.

Note that while the directory structure is logically organized as a hierarchy, a specific Web page may occur in many different parts of the hierarchy. There is usually more than one way to reach a given page.

5.2.2 Popular Directories

- *AOL NetFind*—`www.aol.com/netfind`
- *CNET Search.com*—`www.search.com`
- *Excite*—`www.excite.com`
- *Infoseek*—`www.infoseek.com`
- *Looksmart*—`www.looksmart.com/x02`
- *Lycos*—`www.lycos.com`
- *Magellan*—`www.mckinley.com`
- *Yahoo!*—`www.yahoo.com`

Notice that the URLs for many of the directories have the following form[1]:

`http://www.directoryname.com/`

URLs for search engines, metasearch engines, and white pages often have this form, as well. Some browsers allow you to type just the name of the search tool into the location area. For example, if you enter `infoseek` into the location area in the Netscape browser window and press the return key, the browser expands `infoseek` into

`http://www.infoseek.com/`

and takes you to the `Infoseek` directory.

5.2.3 Search Engines

The second approach to organizing information and locating information on the Web is a search engine, which is a computer program that does the

---

[1] Recall that we omit the leading `http://` and the trailing `/` on URLs, unless discussing a specific point about naming.

following:

1. Allows you to submit a form containing a *query* that consists of a word or phrase describing the specific information you are trying to locate on the Web.

2. Searches its database to try to match your query.

3. Collates and returns a list of clickable URLs containing presentations that *match* your query; the list is usually ordered, with the better matches appearing at the top.

4. Permits you to revise and resubmit a query.

A number of search engines also provide URLs for related or suggested topics.

Many people find that search engines are not as easy to use as directories. To use a search engine, you supply a query by entering information into a field on the screen. To be effective, that is, to have the search engine return a small list of URLs on your topic of interest, you often need to be very specific. To pose such queries, you must learn the *query syntax* of the search engine with which you are working. Learning the syntax so that you can phrase effective and legal queries often requires that you read and understand the documentation accompanying the search engine. A hyperlink to the documentation is usually provided next to the query field, and example queries are often given.

Once you learn to use a specific search engine's query language effectively, you can quickly zoom in on very narrow topics. This is the advantage of a search engine. The disadvantages are that you have to learn the query language and you have to learn a search strategy.

The user-friendliness and power of query languages vary from search engine to search engine. We recommend you try several of them and then learn the syntax of one search engine's query language. Since each search engine searches a different database, you would be best off learning about a search engine that has indexed an extensive amount of material of interest to you. You may be able to gauge this by posing similar queries to a number of search engines and seeing which one finds the best matches.

Some versions of Netscape 4.x allow you to enter a search query directly into the browser's location area. The query must contain a space; otherwise, the browser will try to interpret it as a URL. One of the search engines that Netscape knows about will be used to process the query. For example, typing

```
Killface cat
```

into the browser's location area will result in the query "Killface cat" being posed to a search engine. This saves you a step, because you do not have to bring up a search engine separately before posing a query.

5.2.4 Popular Search Engines

- *AOL NetFind*—`www.aol.com/netfind`
- *AltaVista*—`altavista.digital.com`
- *Excite*—`www.excite.com`
- *HotBot*—`www.hotbot.com`
- *Infoseek*—`www.infoseek.com`
- *Lycos*—`www.lycos.com`
- *Magellan*—`www.mckinley.com`
- *WebCrawler*—`www.webcrawler.com`

By comparing the sites listed in Sections 5.2.2 and 5.2.4, you can see that a number of sites provide directory, as well as search engine, capabilities. This is not surprising, since a large database can support either mechanism if the appropriate interface is in place.

The distinction between directories and search engines may become blurred over time, and a new hybrid category called "direct search" or something similar may be necessary to classify the new search tools.

5.2.5 Metasearch Engines

A *metasearch engine* or *all-in-one search engine* performs a search by calling on more than one other search engine to do the actual work. The results are collated, duplicate retrievals are eliminated, and the results are ranked according to how well they match your query. You are then presented with a list of URLs.

The advantage of a metasearch engine is that you can access a number of different search engines with a single query. The disadvantage is that you will often have a high noise-to-signal ratio; that is, a lot of the "matches" will not be of interest to you. This means you will need to spend more time evaluating the results and deciding which hyperlinks to follow.

For very specific, hard-to-locate topics, metasearch engines can often be a good starting point. For example, if you try to locate a topic using your favorite search engine, but fail to turn up anything useful, you may want to query a metasearch engine.

5.2.6 Popular Metasearch Engines

- *Metasearch*—`www.metasearch.com`
- *Metacrawler*—`www.metacrawler.com`

ELLEN AND RAY'S CHOICES

We have found the following tools to be extremely helpful:

- Directory: *Yahoo!*—www.yahoo.com

- Search engines: *AltaVista*—altavista.digital.com, and *Infoseek*—www.infoseek.com

- Metasearch engine: *Metacrawler*—www.metacrawler.com

- White pages: *Bigfoot*—www.bigfoot.com

- *MetaFind*—www.metafind.com

- *SavvySearch*—guaraldi.cs.colostate.edu:2000

5.2.7 White Pages

White pages provide you with an on-line mechanism for looking up information about individuals. They can be used to track down telephone numbers, s-mail addresses, and e-mail addresses. Users can request that information about them be put into a database for a given set of white pages. Some white pages also permit you to submit requests to delete your listing from their information. Many of the white pages are very comprehensive, listing information about people who are not even listed in the telephone book (including their telephone numbers).

When you submit a request for information about a person, some of the white page services may make a record of your request. In addition, many of the white pages have a hyperlink describing their acceptable use policy. In practice, such policies are impossible to enforce. Obviously, people can abuse white pages, and some people feel that white pages are an invasion of their privacy.

5.2.8 Popular White Pages

- *Bigfoot*—www.bigfoot.com

- *Four11*—www.four11.com

- *WhoWhere*—www.whowhere.com

Yellow pages contain information about businesses.

EXERCISES 5.2 Directories, Search Engines, and Metasearch Engines

1. Pick two directories from Section 5.2.2. How many hyperlinks must you go through to obtain information about the following items: lobster, Toyota

Rav4's, and movie reviews? Based on this limited test, how do the two directories compare?

2. Develop a set of five criteria to compare the directories listed in Section 5.2.2. Try to suggest comparisons that distinguish between each pair of directories.

3. (This exercise is designed to illustrate the differences between search engines' databases.) Search for the word "flamenco" using three search engines mentioned in Section 5.2.4. How many matches did each search return? Did any of the searches retrieve any of the same hyperlinks?

4. Develop a set of five criteria to compare the search engines listed in Section 5.2.4. Try to suggest comparisons that distinguish between each pair of search engines.

5. Design an experiment to compare the speeds of the metasearch engines mentioned in Section 5.2.6. Explain your methodology and findings.

6. How many different search engines does each metasearch engine in Section 5.2.6 call on? Explain why a metasearch engine needs to obtain a balance between too few and too many search engines.

7. Which metasearch engine would you rate as having the most user-friendly interface and why? Is there anything you would do differently?

8. Search for your own name in the white pages listed in Section 5.2.8. What were the results? How many people can you locate in the U. S. with the same name as you?

9. Give the URLs of two yellow page services. Search for a company at which a friend works. Were you successful?

10. Using the yellow pages, can you locate the business, Meeting House Data Communications? Using other Web search mechanisms, can you locate this company? Report your findings.

5.3 Search Fundamentals

Figure 5.1 shows the Infoseek search engine user interface, which we will use to explain search fundamentals and to fill in a few details about what directories and search engines typically display on their Web pages. Although not all search tools have the same features as Infoseek, you should be able to use this discussion as a basis for most of them.

First, notice that there is a lot of information on Infoseek's page. The page also serves as a directory, with a large number of clickable categories.

The page is essentially divided up into six parts: header, information bar, search form area, directory area, Infoseek links, and footer. The header contains the Infoseek logo and some advertising. The information bar, located

FIGURE 5.1 The User Interface of the Infoseek Search Engine (www.infoseek.com).

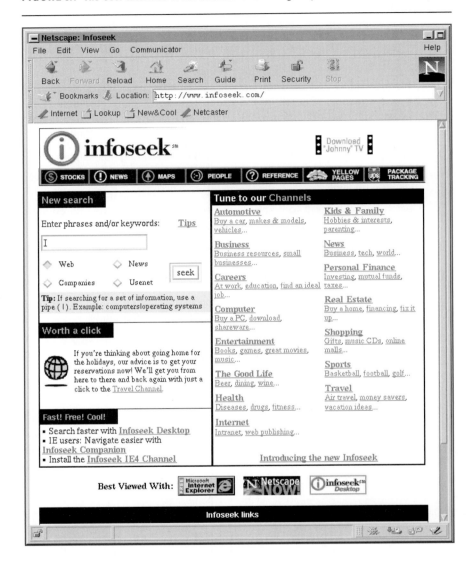

under the header, contains a number of hyperlinks for such services as yellow pages, the United Parcel Service, a news center, and so on. The search form area is located on the left side of the screen, under the information bar, and consists of a form with a field in which you can type a query. You can enter text in the form and then ask the search engine to look for (seek) Web pages that contain information about that text. To the right of the form is Infoseek's directory, with a large number of categories, or channels, displayed as hyperlinks. Certain channels are promoted in the section called "Worth a click," located in the area under the search form in Figure 5.1, while Infoseek software enhancements are offered in the "Fast! Free! Cool!" section.

The section title "Infoseek links" is visible at the very bottom of the screen in Figure 5.1. Various Infoseek links are listed here (but are not visible in the figure) including a hyperlink to jobs available at Infoseek and a hyperlink providing information about advertising using Infoseek. Two other interesting links are provided here and are worth mentioning as well. One is the Add URL hyperlink that allows you to inform Infoseek about a new presentation. The other is the Help hyperlink which contains a wealth of information about Infoseek's search engine. Most search tools contain hyperlinks that serve similar functions.

Figure 5.2 displays the page loaded when Infoseek's Help hyperlink is followed. This page provides everything you need to know to become an effective Infoseek search engine user. It is worthwhile to spend some time familiarizing yourself with this page. One reason we like Infoseek is that its on-line documentation is complete, user-friendly, and well-stocked with good examples. You will learn more about Infoseek's query syntax later, when we discuss various types of search queries.

Lastly, the footer is displayed at the very bottom of the page and is not visible in Figure 5.1. The footer contains information about Infoseek, copyright information, and a *disclaimer*, among other things. The disclaimer contains a statement about Infoseek's "Limitation of Liability" and essentially states that Infoseek does not guarantee the accuracy of any information returned by its search tools. As you will see, all search tools display similar disclaimers.

5.3.1 Search Terminology

Here are a few common search-related terms you should know about.

- **Search tool**   Any mechanism for locating information on the Web; usually refers to a search or metasearch engine, or a directory.

- **Query**   Information entered into a form on a search engine's Web page that describes the information being sought. Note that a query is not usually phrased as a question.

- **Query syntax**   A set of rules describing what constitutes a legal query. On some search engines, special symbols may be used in a query.

- **Query semantics**   A set of rules that defines the meaning of a query.

- **Hit**   A URL that a search engine returns in response to a query.

- **Match**   A synonym for hit.

- **Relevancy score**   A value that indicates how close a match a URL was to a query; usually expressed as a value from 1 to 100, with the higher score meaning more relevant.

FIGURE 5.2 **The On-line Help Provided by the Infoseek Search Engine**
(`www.infoseek.com/Help?pg=HomeHelp.html`).

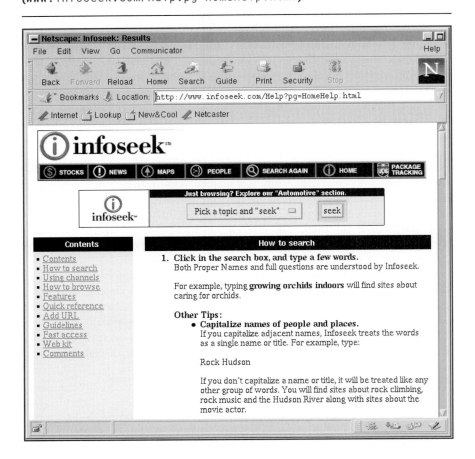

5.3.2 Pattern Matching Queries

The most basic type of query is a *pattern matching query*. You formulate a pattern matching query using a keyword or a group of keywords. The search engine returns the URL of any page that contains these keywords. What we really mean by the phrase "contains these keywords" varies between search engines. For example, some search engines return the URL of any Web page they know about in which the keywords occur. Others return the URL of any page in which the keywords appear within the first 100 words of the page. Still others return only the URLs of those pages in which the keywords appear in the title. Is it necessary that all the keywords appear? Usually, you have the option of specifying whether you want all keywords to be found or whether just one will suffice. The exact details of how pattern matching queries are resolved is search-engine specific.

Suppose you want to find any Web presentations with information about college rankings. A pattern matching query of `College Rankings` submitted to Infoseek turned up 244 hits.[2] This means that Infoseek located 244 pages that contained either the word `College` or the word `Rankings`. Interestingly, a search of `College Ranking` (without the "s") turned up 375 hits. Some search engines perform a process called *stemming*, which means they use variations on the endings of words in your query. For example, the search engine may try the plural form of words, as well as the singular, or vice versa. It appears from our experiment that Infoseek does not use stemming. A search of `college rankings` (no capital letters) turned up 1,763,673 hits. This means that Infoseek knows about close to two million pages that contain either the word `college` or the word `ranking`. We should note that all the searches we conducted with Infoseek required less than ten seconds, even when an enormous number of hits was returned.

Some search engines allow you to specify that keywords must appear next to each other on the Web page. On Infoseek, the query `College-Ranking` means a document must contain these two words together; this query resulted in 377 hits. Also, the query "`College Ranking`" with the quotes, means the same thing as the query `College-Ranking`. The query `College-Rankings` (with the "s") resulted in 246 hits. The query `+College +Ranking` turned up 119 hits; this means there were 119 pages containing both the word `College` and the word `Ranking`. A + sign before a word (or phrase) in an Infoseek query means that the word (or phrase) must occur in the pages to be returned. Clearly, something odd happened in this final search: if 377 pages contain the words `College` and `Ranking` adjacent to each other, then at least this many contain the word `College` and the word `Ranking`.

Table 5.1 summarizes the search results presented here, as well as some other searches. The table illustrates that search engines sometimes exhibit unpredictable behavior. For example, the queries `Rankings College` and `College Rankings` turned up 74 and 244 hits, respectively. According to the Infoseek documentation, these queries should return the URL of any Web page containing either `College` or `Rankings`; therefore, the queries should have returned the same number of hits. Table 5.1 also shows the sensitivity of a search engine to queries with only minor differences—`College Ranking` turned up 375 hits, whereas `college ranking` turned up 1,763,673 hits.

It is important to note that keywords in a query must be spelled correctly. Misspelling a keyword might result in zero hits or hits that are unrelated to your topic.

---

[2] Your search results may vary from ours since the Web and search engines change so quickly. All of the search statistics reported in this section were compiled from searches conducted within a half-hour time period.

TABLE 5.1 Queries Posed to the Infoseek Search Engine and the Resulting Number of Hits.

Infoseek Query	Number of Hits
`rankings-college`	74
`Rankings College`	74
`+College +Ranking`	119
`Ranking College`	128
`"College" "Rankings"`	234
`College Rankings`	244
`college-rankings`	244
`College-rankings`	244
`college-Rankings`	244
`College-Rankings`	246
`"College Ranking"`	371
`College Ranking`	375
`College-Ranking`	377
`college-ranking`	377
`college ranking`	1,763,673
`college rankings`	1,763,676

5.3.3 Boolean Queries

George Boole was a famous mathematician who worked on algebra; *Boolean algebra* was named after him. *Boolean queries* involve the Boolean operations AND,[3] OR, and NOT. Most search engines allow you to enter Boolean queries.

Let us consider some examples of Boolean queries (posed to a generic search engine[4]) to illustrate how they work. A query such as

`paint AND house`

will turn up all Web pages that contain both `paint` and `house`.

Some search engines permit using multiple ANDs. For example,

`Janet AND Tito AND Michael AND LaToya`

---

[3] We use the small capital font to denote Boolean operations. However, when you type a query to a search engine, all the words will appear in the same font.

[4] Our discussion here assumes that the generic search engine *indexes* the full text of a Web page, which is now very common.

would turn up any Web page containing all four of these names. Notice that we have capitalized the names. When posing queries, it is a good idea to capitalize proper nouns and names.

In contrast, the query

```
Husky OR Akita
```

will find all Web pages that contain at least one of the words Husky or Akita; that is, they must contain either Husky, Akita, or both. Analogous to the use of AND, some search engines also permit using more than one OR. In a pattern matching query, the default of some search engines is to OR the words together (meaning find pages containing at least one of the words); for other search engines, the default is to AND the words together (meaning find pages containing all of the words).

To exclude an item from a search, you can use the Boolean NOT operation. For example, to find all Web pages containing information about John Lennon but not the Beatles, a query such as

```
John Lennon NOT The Beatles
```

could be used. To find information that also does not include Yoko, you could try a query such as

```
John Lennon NOT The Beatles NOT Yoko Ono
```

In many search engines, using quotes around a phrase means the words must appear together, in the order you typed them. So, the previous query might be entered as

```
"John Lennon" NOT "The Beatles" NOT "Yoko Ono"
```

The exact syntax of a query will vary from search engine to search engine. For example, some use AND, OR, and NOT, while others use + for AND, nothing for OR (that is, just list the words, which are OR'd by default), and - for NOT.

5.3.4   Search Domain

Most search tools provide some flexibility in the choice of domains to search. For example, you can search the Web, newsgroups, specialized databases, or the Internet. Depending on the item for which you are looking, you may decide to try either a more specific domain first, in hopes of a more efficient search, or a comprehensive and more time-consuming search.

5.3.5 Search Subjects

Several search and metasearch engines provide a way for you to view in real time the search queries of anonymous users. You will get either a ticker tape of queries or a list of URLs that were returned in response to user queries. This is interesting for several reasons:

1. You can see how busy the search tools are.

2. You can "spy" on other anonymous users.

3. You can sometimes "see" the same user modifying their search. For example, you may see a typo corrected in a subsequent query.

4. You can see the variety of people's interests.

5. You may turn up a page of personal interest that you otherwise might not have come across.

EXERCISES 5.3 Search Fundamentals

11.   Repeat the following queries three times, using three different search engines (nine queries in total):

(a) World's Highest Mountain.

(b) Mount Everest.

(c) Everest.

What were your results? Explain.

12.   Execute a Boolean query on your favorite search engine to locate information about Lewis and Clark.

13.   On your favorite search engine, what query would you use first to find information about appetizers and desserts? Experiment with other types of queries. Which query yielded the best results?

14.   Phrase a query to locate information about Janet Evans and swimming, but does not include any reference to Stanford. What were your results?

15.   Phrase a query about cycling that does not retrieve any information about mountain biking. Experiment and explain the results. What was your most productive query?

16.   (The purpose of this exercise is to help you determine how many pages a given search engine has indexed.) Try to maximize the number of hits you are able to get in a single query to a search engine. What was the most hits you were able to get, what was the query, and what search engine did you use? Repeat this problem using a metasearch engine.

17.   Perform the following pattern matching queries, using a single search engine: `waffle`, `Waffle`, `Belgian Waffle`, `waffles`, `Waffles`, and `Belgian Waffles`. What were the resulting numbers of hits? Were they as expected? Explain.

5.4 Search Strategies

Determining which search engine to use can be challenging. You can begin by testing a number of different search engines, trying to find one that you believe meets the following conditions:

- Possesses a user-friendly interface.

- Has easy-to-understand, comprehensive documentation.

- Is convenient to access; that is, you do not have to wait several minutes before being able to submit a query.

- Contains a large database, so that it knows a lot about the information for which you are searching.

- Does a good job in assigning relevancy scores.

If you can find a search engine that meets most of these criteria, you should concentrate on learning it well, rather than learning a little bit about several different search engines.

Once you have learned the query syntax of that search engine, you can begin to formulate your search strategy. When you post queries to the search engine, two common situations can occur: either your query does not turn up a sufficient number of hits, or your query turns up too many hits. In the next two sections, you will learn strategies for dealing with these situations.

5.4.1 Too Few Hits: Search Generalization

Suppose your query returns no hits or only a couple of hits, neither of which is very useful to you. In this case, you need to generalize your search. The ways to do this include:

- If you used a pattern matching query, eliminate one of the more specific keywords from your query.

- If you used a Boolean query, remove one of the keywords or phrases with which you used AND, or delete a NOT item you specified.

- If you restricted your search domain, enlarge it.

- If you are still having no luck, try keywords that are more general, or exchange a couple of the keywords with synonyms.

- If this fails, you may decide to use a directory and work your way down to the topic of interest. Another alternative would be to use a metasearch engine.

5.4.2 Too Many Hits: Search Specialization

Suppose your query returns more URLs than you could possibly look through. In this case, you need to specialize your search.

- If you started with a pattern matching query, you may want to add more keywords.

- If you began with a Boolean query, you might want to AND another keyword, or use the NOT operator to exclude some pages. When you phrase a new query, many search engines let you search just the results of your original search or refine your search. This restricts the domain of your search. What you are essentially doing in this case is a search within a search.

- If you are still retrieving too many hits, try capitalizing proper nouns or names.

- If nothing seems to work, try reviewing the first 20 or so URLs, since search engines list the best matches near the top. If they do not contain what you are looking for, the information they do contain may help you refine your search.

- If this fails, you could resort to a directory and work your way down to the topic of interest.

5.4.3 Sample Searches

Suppose you are interested in sea kayaking through Prince William Sound in Alaska. What is a good search strategy for locating information about this on the Web? We will explore a number of different search queries, using Infoseek, to illustrate how this might be done. The process carries over to other searches.

One of the keys to an effective search strategy is to examine the results you obtain and then revise your search accordingly. Initially, we perform a very general search using the keyword alaska. This turns up 176,954 hits. Next we decide try Alaska, but even this turns up 176,064 hits, which is still far too many to wade through. At this point, it is obvious the search is too general. We next decide to enter the query

```
+"Prince William Sound" +Alaska
```

Recall that the quotes around Prince William Sound indicate a phrase, and the + symbols specify that both "Prince William Sound" and Alaska must appear. This results in 778 hits; however, two of the first three of these contain the word kayaking in their titles. Normally, we would probably explore these URLs, but for illustration purposes, we will continue our search.

Next, we decide to include the word `kayak` in the query:

```
+kayak +"Prince William Sound" +Alaska
```

This search results in 44 hits, with nearly every title containing the word kayak. Since we are really interested in kayaking, let us change `kayak` to `kayaking`:

```
+kayaking +"Prince William Sound" +Alaska
```

The query results in 60 hits, but in one sense, we have lost a little ground. (This is typical of a search.) We know we need to rent the kayaks while in Alaska, so let us add this concept into the search:

```
+kayaking +"Prince William Sound" +Alaska +rental
```

This search results in 20 hits. At this point, we are comfortable searching through the 20 URLs to locate the available information on kayaking in Prince William Sound. An experienced user might have attempted a query similar to our final query to begin with. Table 5.2 summarizes the queries.

Once you learn to use a search engine well, most of the queries you pose will get you closer to your goal. Very few of the queries will be wasted or will cause you to lose ground.

TABLE 5.2 Queries about Kayaking in Alaska, Posed to the Infoseek Search Engine, and the Resulting Number of Hits.

Infoseek Query	Number of Hits
alaska	176,954
Alaska	176,064
+"Prince William Sound" +Alaska	778
+kayak +"Prince William Sound" +Alaska	44
+kayaking +"Prince William Sound" +Alaska	60
+kayaking +"Prince William Sound" +Alaska +rental	20

18. A friend of yours is new to the Web and search engines. Write one paragraph that explains to them how to generalize a search and another that explains how to specialize a search.

19. In traversing through a directory to locate information about kayaking in Prince William Sound, how may different URLs did you have to follow? What were the corresponding categories you went through?

20. Your parents are interested in buying a VCR from Sony. Can you locate any information on the Web from *Consumer Reports* about Sony VCRs? Describe your search strategy.

21. Try to locate ticket information for the next Olympic Games. Were you able to find out what ticket prices are going to be? What queries did you perform, and what were the results of those queries?

5.5 How Does a Search Engine Work?

5.5.1 Search Engine Components

If you understand how a search tool works, there is a good chance you will be able to use it more effectively. In this section, we describe how a search engine works. For the most part, these same ideas apply to directories; the main difference is that the hierarchical organizational structure and categorizations for directories need to be in place and displayed. The references include additional information about how directories are put together.

To describe how a search engine works, we split up its functions into a number of components: user interface, searcher, and evaluator.

User interface   The screen in which you type a query and which displays the search results.

Searcher   The part that searches a database for information to match your query.

Evaluator   The function that assigns relevancy scores to the information retrieved.

In addition, a search engine's database is created using the following:

Gatherer   The component that traverses the Web, collecting information about pages.

Indexer   The function that categorizes the data obtained by the gatherer.

For comparison, think of the different facets of a typical library, such as acquisitions, cataloging, indexing, and on-line searching.

5.5.2   User Interface

Figure 5.1 depicted the Infoseek user interface, which is typical for a search engine. The user interface must provide a mechanism by which a user can submit queries to the search engine. This is universally done using forms. In addition, the user interface should be friendly and visually appealing. Hyperlinks to `help` files should be displayed prominently, and advertisements should not hinder a reader's use of the search engine. Finally, the user interface needs to display the results of the search in a convenient way. The user should be presented with a list of hits from their search, a *relevancy score* for each hit, and a "summary" of each page that was matched. This way, the user can make an informed choice as to which hyperlinks to follow.

5.5.3   Searcher

The *searcher* is a program that uses the search engine's index and database to see if any matches can be found for the query. Your query must first be transformed into a syntax that the searcher can process. Since the databases associated with search engines are extremely large (with perhaps 25,000,000–50,000,000 indexed pages), a highly efficient search strategy must be applied. Computer scientists have spent years developing efficient search and sorting strategies; some of these sophisticated algorithms are implemented in the searcher. More details about the basic principles behind these strategies can be found in any introductory computer science algorithms textbook.

5.5.4   Evaluator

The searcher locates any URLs that match your query. The hits retrieved by your query are called the *result set* of the search. Not all of the hits will match your query equally well. For example, a query about "Honey Bees," might be matched by a page containing the phrase "Honey Bees" in the following sentence:

```
Ants, honey bees, and crickets are all insects.
```

or by the page titled

```
Everything You Ever Wanted to Know About Honey Bees
```

Clearly, in most cases, it would be better to rank this second page much higher, as it probably contains many more references to Honey Bees. The ranking process is carried out by the evaluator, a program that assigns a relevancy score to each page in the result set. The relevancy score is an indication of how well a given page matched your query.

How is the relevancy score computed by the evaluator? This varies from search engine to search engine. A number of different factors are involved, and each one contributes a different percentage (according to a weighting scheme) towards the overall ranking of a page. Some of the factors typically considered are:

- How many times the words in the query appear in the page.

- Whether or not the query words appear in the title.

- The proximity of the query words to the beginning of the page.

- Whether the query words appear in the CONTENT attribute of the meta tag.

- How many of the query words appear in the document.

Some search engines also consider other factors in computing a relevancy score. Each factor is weighted, and a value is computed that rates the page. The values are usually normalized and are assigned numbers between 1 and 100, with 100 representing the best possible match. As part of the user interface, the result set and relevancy scores computed by the evaluator are displayed for the user, with the best matches appearing first. Hyperlinks to each hit are provided, and a short description (two lines or so) of the page is usually given.

With many search engines, you can set the maximum number of hits that you want returned. This can make your search more efficient, since fewer hits need to be found and ranked. Once the actual number of hits is displayed and you page through them based on their relevancy scores, you can decide whether to enlarge your search.

The algorithms used by evaluators are imperfect, but they are getting better. Also, in addition to direct hits, a search engine will sometimes display the hyperlinks of pages that contain information about related topics.

5.5.5 Gatherer

A search engine obtains its information by using a *gatherer*, a program that traverses the Web and collects information about Web documents. The gatherer does not collect the information every time a query is made. Rather, the gatherer is run at regular (short) intervals, and it returns information that is incorporated into the search engine's database and is indexed. Alternate names for gatherer are *bot*, *crawler*, *robot*, *spider*, and *worm*.

A gatherer may employ essentially two different methods to search the Web for new pages. (In practice, hybrids of these procedures are often used.) Both techniques are well-known search strategies in computer science; they are called *breadth-first search* and *depth-first search*. We will describe both methods on a basic level. The interested reader may consult any introductory computer science algorithms textbook for a more detailed explanation.

Breadth-First Search

A breadth-first search proceeds in levels "across" the pages. The gatherer begins at a particular Web page and then explores all pages that it can reach by using only one hyperlink from the original page. Once it has exhausted all Web pages at that one level, it explores all of the Web pages that can be reached by following only one hyperlink from any page that was discovered at level one. In this way, a second level, which usually contains many more Web pages than the first level, is explored. This process is repeated level by level until no new Web pages are found. When no more pages can be located, the search may need to jump to a new starting point.

Figure 5.3 illustrates the process. In the figure, we have greatly simplified things by representing a very small corner of the Web as two discrete collections of documents. Each Web page is represented by a circle, and hyperlinks

FIGURE 5.3 An Illustration of the Order in Which a Gatherer Visits Web Pages Using a Breadth-First Search Strategy.

The primary number indicates the order in which the page was visited, and the subscript indicates the level in which the page was found. The document collection used here is a simplified one.

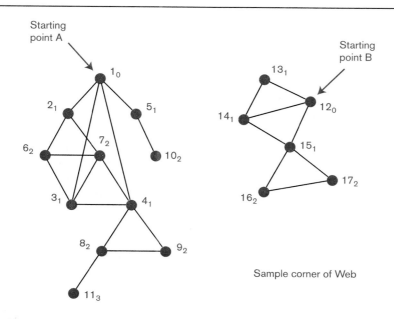

are represented by lines connecting the circles. We assume that hyperlinks go in both directions (another simplification). The search begins from point A. We have numbered the Web pages in the order that they are visited by our hypothetical gatherer.

The initial Web page where the search begins is labeled 1_0. The subscript shows the level in which the page appears. The gatherer will normally choose a random hyperlink to follow from its current level. For illustration purposes, we have always chosen to follow the leftmost link first. So, for example, the first page explored from page 1_0 is its leftmost hyperlinked page, which is labeled 2_1. This is the second page visited, and it appears at level one. The three other pages discovered at level one are 3, 4, and 5. Pages 6, 7, 8, 9, and 10 are found at level two, and page 11 is found at level three.

Having found page 11, the gatherer has exhausted the entire collection of Web pages hyperlinked to the starting point A. The gatherer then chooses a new starting point B and continues from there. As before, we begin with level zero. Details such as how new starting points are chosen, how the gatherer knows it has been somewhere already and so does not go into an infinite search loop, how the gatherer reports its findings, and so on, are not important for our discussion. The curious reader should consult the references for further information about these issues.

Depth-First Search

A depth-first search proceeds by following a chain of hyperlinks "down" as far as possible. The gatherer begins at a particular Web page and explores one of its hyperlinks. At the new page, the gatherer follows another hyperlink. At the next page, one of its hyperlinks is followed, and so on. In contrast to the breadth-first search, hyperlinks on a given page are not fully exhausted before the gatherer goes to the next level page. When the gatherer reaches a page from which no new pages can be discovered, the search backtracks until it can go forward again and discover new pages. The search goes as deep into the document collection as possible before backtracking. As with a breadth-first search, when no more pages can be located, even with full backtracking, the search jumps to a new starting point.

Figure 5.4 depicts the depth-first search process. For illustration purposes, we use the same document collection as we did for the breadth-first search example. The search begins at point A. We have labeled the Web pages in the order they are visited by our hypothetical gatherer. The initial Web page where the search begins is labeled 1. The gatherer will normally choose a random hyperlink to follow from its current location. However, as we did in the breadth-first case, we have always chosen to follow the leftmost link first. So, for example, the first page explored from page 1 is its leftmost hyperlinked page, which is labeled 2 as the second page visited. With a depth-first search, the levels on which the different pages are found are not as important as in breadth-first search.

FIGURE 5.4 An Illustration of the Order in Which a Gatherer Visits Web Pages Using a Depth-First Search Strategy.

Each number indicates the order in which the page was visited. The document collection used here is the same as that used in Figure 5.3.

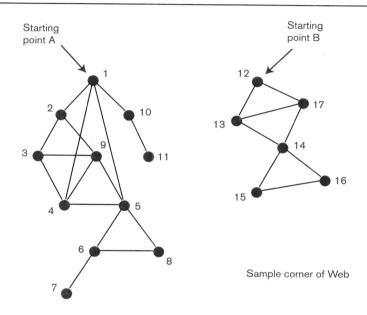

Notice that a deep chain is followed from page 1, reaching pages 2, 3, 4, 5, 6, and 7 before the search needs to backtrack. The search backtracks and discovers page 8 from page 6. Again, the search must backtrack and page 9 is discovered from page 5. After backtracking once more, the gatherer located page 10 from page 1 and page 11 from page 10. Having found page 11, the gatherer has exhausted the entire collection of Web pages hyperlinked to the starting point A. This is not discovered until we have backtracked to page 1. A new starting point B is chosen and the gathering continues from there. For illustration purposes, we continue the numbering from where we left off. Pages 13, 14, 15, and 16 are found by following a single chain from page 12. Backtracking to page 14, we find page 17. Once the search backtracks to page 12, we determine that the entire document collection reachable from starting point B has been exhausted.

Precise details of how a depth-first search is carried out are not important for our discussion. Curious readers should consult the references.

Miscellaneous Facts about Gatherers

- Gatherers place a very heavy load on Web servers.
- The depth of searches that gatherers perform is restricted.
- Gatherers have trouble dealing with documents that are created using frames.
- The best gathering strategy depends on the underlying collection of documents.
- Some gatherers may only retrieve the head of a document, while others retrieve the entire content of the page. *Full text indexing*, which allows you to search for any word in the entire text of a document, requires that the entire document be retrieved. Full text indexing requires a tremendous amount of storage space.

5.5.6 Indexer

Once the gatherer retrieves information about Web pages, the information is put into a database and *indexed*. The *indexer* function creates a set of keys (an index) that organizes the data, so that high-speed electronic searches can be conducted and the desired information can be located and retrieved quickly.

Libraries have card (or computerized) catalogs that index books by author name(s), by title, and by subject. Each book has a unique ISBN. The equivalent elements that should go into a Web page's bibliographic record include the URL, document title, and descriptive keywords. What else should be included? This question is still being debated. Because resources on the Web vary so widely and because they change so rapidly, indexing is more difficult, and indexes need to be rebuilt frequently. There was a time when about one in three URLs that a search engine returned was out of date. Now that indexes are being rebuilt more frequently and more pages have stabilized, the ratio is not quite as high.

5.5.7 Summary

A search engine's functionality depends on a number of different components: user interface, searcher, evaluator, gatherer, and indexer. The software involved in any single item is very complex, and tying all of the different modules together so that they interface properly is also a complex task.

EXERCISES 5.5 How Does a Search Engine Work?

22.   Is there any search engine for which you can locate information that describes how its specific search strategy works? Does having this information help you in posing more intelligent queries?

23. Define the phrase *Uniform Resource Characteristic (URC)* and describe the role it is designed to play in the indexing of Web pages.

24. Can you locate any information describing how Web pages are indexed by any specific search engine? If so, describe the process in the particular case you found and list your reference(s).

25. How does Infoseek compute its relevancy score?

26. Do you use the max hits option when searching? Explain.

27. Write a paragraph summarizing the items that you feel are the most important in computing a relevancy score.

28. Have you found that the evaluator for a particular search engine is better than that of other search engines? Explain.

29. Do some search engines compute result sets for common queries in advance? That is, are there search engines that compute the answer to an expected query themselves, before the query is asked by any user? If so, what are the tradeoffs in doing this?

30. How does a metasearch engine work?

Telnet and FTP

6

6.1 Introduction

The goals of this chapter are to acquaint you with the following topics:

OBJECTIVES

- Telnet and remote login.
- File transfer.
- Computer viruses.

6.2 Telnet and Remote Login

Telnet and *remote login* are two programs that allow you to log in to another computer from an account in which you are already logged. They let you use and interact with software on the remote machine. To do this, you will need a second computer account that is accessible to you. The second computer is usually at a different physical location; hence the phrase "remote login." Figure 6.1 illustrates the basic idea. Rita is logged in to computer A. Using Telnet or remote login, she logs in to computer B. It is almost as though she were physically at computer B entering commands.

FIGURE 6.1 Rita Was Logged in to Computer A, **and from There, She Remote Logged in to Computer** B.
She is executing commands on computer B **via computer** A.

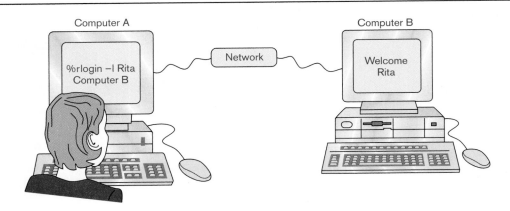

6.2.1 Telnet

The `telnet` command uses the *Telnet protocol* to log in to a remote computer on the Internet. The command is often called `telnet`, but different programs may use other names, such as *tn3270*, *WinQVT*, and *QWS3270*.

A wide range of Telnet clients provide user-friendly interfaces. Most of them function in a similar manner. Here, we describe several different ways of invoking Telnet. On a desktop system, a Telnet client can usually be launched from one of the system's menus simply by selecting the Telnet option. In a Windows environment, the interface may appear as shown in Figure 6.2. Selecting the `Remote System` option from the `Connect` pull-down menu causes the `Connect` window to display within the Telnet window. The form in the `Connect` window (which is visible in Figure 6.2) lets you specify the hostname, port, and terminal type of the computer to which you are connecting. Convenient on-line documentation is available with most Telnet clients, and Figure 6.3 illustrates the help screen available for "Connecting to a Remote Computer."

On some systems, such as UNIX, you can type the command `telnet` at the operating system prompt. When you receive the Telnet prompt,

```
telnet>
```

you can type the `open` command, followed by the hostname of the computer you would like to connect to:

```
telnet>open hostname
```

FIGURE 6.2 A Windows Telnet Screen.

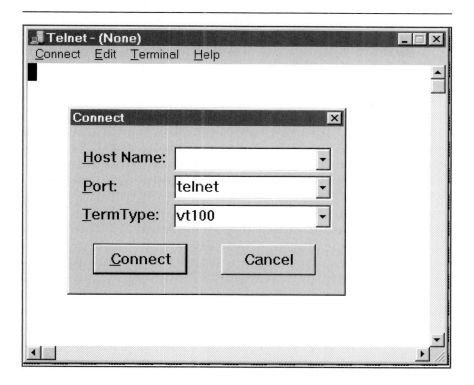

The hostname is the machine domain name (for example, hopper.unh.edu) or the numerical Internet address of the machine. In some cases, you may have to specify a port number, as well. Typing help or ? at the Telnet prompt will usually result in the Telnet documentation being displayed.

If you invoke Telnet on a Mac, the Telnet commands will become part of the Mac's menu bar (as is customary of applications on a Mac). From these menus, you can select the open connection option. This will provide you with a field in which you can enter the name of the remote system.

From within your browser, if you have a Telnet application configured, you can enter a URL such as

```
telnet://hostname
```

and you will be provided with a Telnet window. Figure 6.4 depicts a window that was produced in this manner. In this case, the URL we specified was

```
telnet://hopper.unh.edu
```

FIGURE 6.3 Windows Telnet Help Screen.

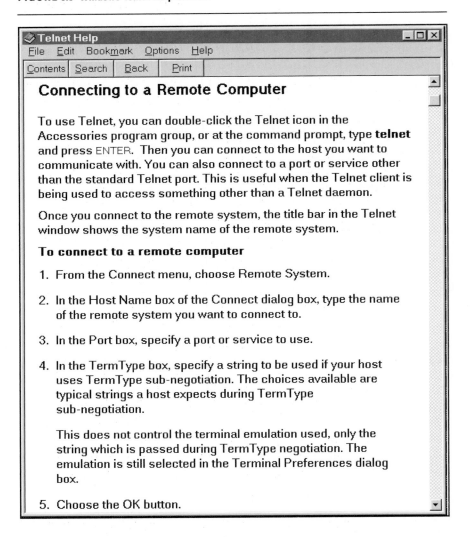

Once you have specified the hostname of the computer and established a connection using Telnet, you can log in as you normally would (assuming you have an account) and begin executing commands on the remote machine.

One important thing to notice when you first bring up Telnet is the *escape sequence* that is displayed on the screen. It can be used to disconnect from the remote machine in case there is a problem. On most systems, the escape sequence is CONTROL-]; that is, by typing CONTROL-] you can disconnect from the remote machine. The CONTROL key must be used because keys typed without CONTROL are interpreted as inputs to the remote machine. Figure 6.4

FIGURE 6.4 A Telnet Window.

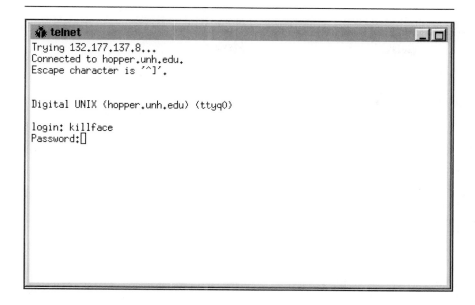

shows the escape sequence message. In the Windows Telnet interface (Figure 6.2), you can also terminate Telnet by selecting `Disconnect` or `Exit` from the `Connect` pull-down menu.

As we mentioned before, it is usually necessary to have an account on the remote system in order to Telnet to it. However, there are some systems, including libraries, that allow "guests" to access their databases (though registration may be necessary). In the on-line portion of this book, we provide information about some of the sites that offer "guest" or free login.

One of the most common uses of Telnet is to log in to your personal machine to retrieve e-mail when you are traveling. Be warned that the process of reading e-mail in this fashion can be very tedious from many countries. The connections are often so slow that it is impossible to retrieve, read, and compose messages. If you are going to travel and if you plan to read e-mail while away, you should probably unsubscribe from all mailing lists you are on.

6.2.2 Remote Login

The `rlogin` command is similar to the `telnet` command, except that it provides the remote computer with information about where you are logging in from. If the machine that you are performing the remote login from is listed in the remote machine's file of hostnames, you may not have to enter a password. On UNIX systems, the list of hostnames is given in a hidden file called `.rhosts`. From a UNIX prompt, the syntax for the `rlogin` command is

```
%rlogin hostname
```

where hostname is the name of the machine from which you want to establish a remote login connection.

As with Telnet, once you are logged in to a remote machine, you will not be able to execute commands using your local login session. All the commands entered will run on the remote machine until the remote session is terminated by using an exit command.

A number of flags can be specified to the UNIX rlogin command. Executing the

```
%man rlogin
```

command will return help about rlogin. By using the -l option, you can start a session on the remote machine with an account name that is different from the one you are currently using on your local machine. For example, suppose Martin McGrath (account name mmg) enters the command

```
%rlogin -l oreilly hal9000
```

He is going to log in to the remote machine hal9000 using the account of his girlfriend, Mary O'Reilly. Of course, to be successful, Martin must know Mary's password.

As a final note, Telnet is a more secure remote login mechanism than rlogin.

EXERCISES 6.2 Telnet and Remote Login

1. Can you Telnet to the machine you regularly use? Can you rlogin to this same machine? If so, is there a reason why you would want to do this?

2. Invoke your Telnet client. What is the escape sequence used on your system?

3. Configure your browser so that you can enter Telnet URLs to launch Telnet. Describe what you needed to do to accomplish this.

4. Research Telnet and remote login using the Web. Write a paragraph comparing and contrasting the security risks of each.

6.3 File Transfer

There are times when you may need to transfer a file from one computer to another. For example,

- You scan in images on one system, and you need to move them to another for permanent display.

- You work on a computer at home, and you need to transfer a file to a machine at your office.

- You and a collaborator need to exchange files.

- You want to download a helper, plug-in, or freeware application from another computer.

- You develop software on one machine and need to move it to another.

File Transfer[1] is an application that allows you to transfer files between two computers on the Internet or on the same network. The two most important facilities provided by file transfer are the abilities to:

1. Copy a file from another computer to your computer.

2. Send a file from your computer to another computer.

Figure 6.5 depicts the intuitive idea of the process. In the figure, fileA resides on computer A. A file transfer connection to computer B is opened. Next fileA is sent over the network to computer B, using file transfer. The figure shows that fileA has arrived at computer B. When copying files, you should first run virus detection software (see next section) on them before using them on your computer. This helps safeguard against your computer getting infected, but it is not a guarantee.

Although file transfer is the fastest and most convenient method, there are other ways of copying a file from one computer to another. For example, if the file to be transferred is not too large, you may be able to e-mail the file to an account on the second machine. In some circumstances, e-mailing a file is not practical, especially when it may be requested many times, as is often the case with freeware. Another option is to copy a file onto a diskette and then use the diskette to install the file on a different machine. However, for computers separated by hundreds or thousands of miles, "walknet" (or "sneakernet") is not a satisfactory solution.

6.3.1 Graphical File Transfer Clients

Graphical file transfer clients are the easiest to use. These applications display the sending computer's file system in one window and the receiving computer's file system in a second window. Figure 6.6 shows a typical user interface. In this scenario, you first need to log in to each computer. This prevents unauthorized users from connecting to a machine and transferring files.

---

[1] We use this as a generic name here. Your file transfer client may have a different name.

FIGURE 6.5 Schematic of How File Transfer Can Be Used to Copy a File from Computer A **to Computer** B.

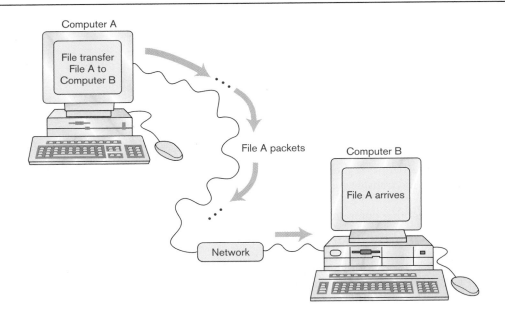

FIGURE 6.6 Windows FTP Client.

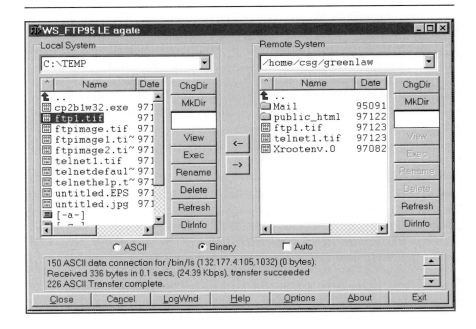

Using a graphical file transfer client is very simple. To transfer a file from one system to another, you can "drag" it using the mouse and "drop" it on the other system. Files can thus be exchanged in either direction.

One important point is the *transfer setting mode*. This can usually be specified by clicking on a button. Most clients have a *text transfer mode* and a *binary transfer mode*. All file types can be transferred correctly using binary mode, but not all file types can be transferred using text mode. Therefore, the mode should be set to binary when transferring images, files containing special characters, or executable files. When in doubt, transfer a file using the binary mode setting. Notice in Figure 6.6 that the binary mode setting is on.

6.3.2   Text-Based File Transfer Clients

There are text-based file transfer clients, as well as graphical systems. For example, you can launch the UNIX file transfer client called *File Transfer Protocol (FTP)* by entering the command

```
%ftp hostname
```

Here `hostname` is the name of the computer with which you want to exchange files. Once you have successfully initiated an FTP session, you will usually be greeted by the prompt `ftp>` or something similar.

At this point, you can enter a variety of commands. Some may look and function like UNIX commands, but others will vary depending on the file transfer program being used. The most important commands allow you to `get` a file (obtain a copy) and `put` a file (deposit a copy). The following list summarizes typical commands for a text-based file transfer client. (The list is not intended to be exhaustive.)

- `bye`   Terminate the session and exit the file transfer program.
- `cd`   Change directory.
- `get`   Copy a file.
- `help`   View a list of commands or help on a specific command.
- `ls`   List the files in the current (or working) directory.
- `put`   Send a copy of a file.
- `pwd`   Print the name of the current directory.

Entering `help` will display a list of commands available at the site, and `help` followed by a command name will provide a description of how that particular command is used. If you are on a UNIX system, entering `man ftp` will provide documentation about FTP.

The command used to retrieve a copy of a file is get. Its syntax is

```
ftp>get file1 file2
```

In this example, the file being retrieved from the remote machine is called file1. A copy of file1 will be placed in the current directory on your local machine, in a file with the name file2. If file2 is omitted, the file you are copying retains the same name, in this case file1. The syntax and semantics of the put command are similar. To end your FTP session, type bye at the prompt.

6.3.3 File Compression

It is common to *compress* files that are to be transferred between two computers. Compressing a file makes it smaller, and the compressed file can be transferred more quickly over a network. A wide variety of compression tools are available, including: *compress*, *gzip*, and *PKZIP*. When you transfer a file, the file extension will usually alert you to whether or not the file is compressed (.Z for compress, .gz for gzip, and .zip for PKZIP). You will need to decompress the file to use it. (On some systems, you can set things up so that files are automatically decompressed for you.) You must use the appropriate decompression tool, which depends on how the file was compressed: *uncompress* for compress, *gunzip* for gzip, and *PKUNZIP* for PKZIP. There are a number of sites from which you can download compression and decompression programs.

File compression is an extensive field in computer science, and the algorithms used are very interesting. Consult an introductory computer science algorithms textbook to learn more about this subject.

6.3.4 Anonymous File Transfer

On some systems, files are made available to anyone who wants to retrieve them (for example, *freeware*, public documents, and so on). If a file needs to be widely distributed, it may not be feasible to assign accounts and passwords to everyone interested in receiving a copy of the file. *Anonymous file transfer* was established in response to this problem. Users log in with anonymous as their account name and then provide a password. The standard practice is to use your e-mail address as a password, because this helps the administrator at the file transfer site monitor file transfer usage. On other systems, a password of guest is expected. You should never use your real account password when logging in anonymously, since the information you submit is not secure.

After logging in via anonymous file transfer, you will be restricted to specific directories. Usually, accessible directories are in an area named something like public. Within those directories, you will often be confined to just downloading files. In many cases, such directories contain a help file, called

FIGURE 6.7 An Illustration of an Anonymous File Transfer Session.

```
nxterm                                                                    _ □
%ftp wonder.german.unm.edu
Connected to wonder.german.unm.edu
220-Department of German
220-University of New Mexico
220-
220-Hello greenlaw@amethyst.cs.unh.edu.
220-
220-We do allow anonymous ftp.  Please use your full email
220-address as a password.
220-
220 wonder FTP server (Version 4.109 Wed Nov 29 09:55:51 MST 1995) ready.
Name: anonymous
331 Please enter your full email address as a password.  Restrictions apply.
Password:
530 Login complete.
Remote system type is UNIX.
Using binary mode to transfer files.
ftp>ls

...

226 Transfer complete.
ftp>get stamp
local: stamp remote: stamp
200 PORT command successful.
150 Opening BINARY mode data connection for stamp (145.162.5.165,30385) (340 bytes
226 Transfer complete.
340 bytes received in 0.0327 secs (10 Kbytes/sec)
ftp>bye
221 Goodbye.


[greenlaw@amethyst SCREEN]$ []
```

something like README, where additional information can be found. Figure 6.7 illustrates an anonymous file transfer session. In this case, after logging in, we checked to see what files were available and then retrieved one of them, called stamp, using the get command.

Note that file transfer can also be launched from within a browser window by entering the URL

```
ftp://hostname
```

Here, hostname specifies the site to which you want to connect.

6.3.5 Archie

Unless you know exactly where a file is located on the Internet, it will be very difficult to find it, since files can be archived at file transfer sites scattered throughout the world. *Archie*, derived from the phrase "file archive," is a service that maintains databases containing listings of files from various file transfer sites. Performing a query to find a specific file involves either sending e-mail to an Archie site or connecting to an Archie server via Telnet. Archie will respond and will indicate the location(s) of the file you want, provided that the file is listed in one of its databases. To locate an Archie server, you can connect to any of the major search engines and enter the query "Archie."

EXERCISES 6.3 File Transfer

5. Do you have a graphical, textual, or both type(s) of file transfer client(s) available on your system? What are the transfer modes you can use with your client(s)?

6. What factors determine how quickly a file transfers over the Internet?

7. Print out a sample file transfer session in which you transfer a file to your system.

8. What file compression programs are available on your system? Can they be configured to decompress compressed files automatically?

9. Try to compress several different types of files. What space savings did you obtain? Did the file type make a difference?

10. Does file compression or file decompression take longer? Experiment. Explain your findings.

11. Research Archie using the Web. Summarize in a paragraph what you learned.

12. Provide the addresses of two "nearby" Archie servers. Try to connect to one of them and locate file compression programs. Were you successful?

13. What is *Veronica*? Conduct research on the Web and describe your findings.

6.4 Computer Viruses

Some of the programs that you download from the Internet, or obtain as e-mail attachments may threaten the security of your computer if they contain *viruses*, *Trojan horse programs*, or *worms*. In this section, you will learn different strategies for handling such annoyances.

6.4.1 Definitions

A virus can be thought of as a program that, when run, is able to "replicate" and then embed itself within another program. Although there are harmless viruses, most are intended to do damage to the host system. The damage can occur immediately, by filling all available space on your hard drive, for instance, or it may occur at some later time, after the virus has had a chance to be passed along to other programs and computers. The damage to your computer might involve something as innocent as a message being displayed on your desktop, or, more likely, the destruction or modification of data or the deletion of files. Before doing the damage, however, the virus could infect other programs on your computer, as well as other computers if you send program files to anyone else and they run them. A specific event, such as the tenth time that the host program is run, or a particular date on which the host program is run, may *trigger* the virus to become active.

A Trojan horse program (the name comes from Greek mythology) is a legitimate program for carrying out some useful function, but within it is hidden code that is activated by some trigger. When the hidden code is executed, it might release a virus, permit unauthorized access to the computer, or destroy files and data.

A worm is a stand-alone program that tries to gain access to computer systems via networks. For example, a worm might try various password combinations until it is successful. The 1988 *Internet Worm*, created by Robert T. Morris, is a highly successful example. Although not designed to be destructive (the worm was intended to be an experiment), the worm caused major problems when it inadvertently consumed the available memory in the systems it invaded.

6.4.2 Virus Avoidance and Precautions

Relatively speaking, UNIX viruses are rare, because of the strict security measures on UNIX systems. Most viruses are designed to infect either PCs or Macs. A virus is usually targeted at one type or the other of such systems, since nearly all viruses are operating-system specific. To protect yourself from viruses and Trojan horse programs, we recommend the following precautions:

- Run *antivirus software* (also called *virus detection software*) on any new programs. This software looks for viruses and Trojan horse programs by comparing data patterns found in your programs to characteristic data patterns found in programs infected by known viruses.

- Do not download files from unknown sources. This includes mail attachments from individuals and organizations unknown to you. If you do download something, run antivirus software on it before opening the file or running the program.

- Do not use pirated copies of software.

- Keep your antivirus software up to date, since new releases will contain information necessary to identify the latest viruses and Trojan horse programs.

- Back up your files regularly (after ensuring that they are virus-free by running antivirus software on them). If you do lose files or data because of a virus, you will be able to recover if you have current backup files.

Although viruses, Trojan horse programs, and worms are a very real computer security threat, they also present an opportunity for hoaxes and misinformation. A common misunderstanding is that viruses can somehow be transmitted directly via e-mail. This was the premise of the "Good Times" hoax, a message that stated there was a virus being spread through e-mail. The alleged viral e-mail message contained the expression "Good Times" in the subject, and recipients were warned not to download or even read the message. They were then urged to forward the hoax message to everyone they knew. Of course, a regular e-mail message is a text file, and viruses need to be embedded in programs, so reading an e-mail message is fine; however, downloading and running programs that are attachments to e-mail require caution. An interesting Web page dedicated to Internet hoaxes is the CIAC Internet Hoaxes Page, and a link is provided in this book's accompanying Web presentation.

EXERCISES 6.4 Computer Viruses

14. Research the Microsoft Word Macro Viruses. How many different strains are there? Were you surprised by this? Why is it so easy to have viruses in Word? Speculate why these viruses spread so rapidly.

15. Have you or a friend ever had problems with computer viruses? Describe your experiences. Do you regularly use virus detection software? Did you receive the "Good Times" message? When? Did you forward it?

16. How can you obtain virus detection software? Explain how to get around the danger of obtaining virus detection software that is itself infected.

7

7.1 Introduction

In this chapter, we expand our discussion of HTML, covering the following topics:

OBJECTIVES

- Semantic and syntactic style types.
- Headers and footers.
- Lists.
- Tables.
- Debugging.

7.2 Semantic Versus Syntactic Based Style Types

A large number of HTML tags are used for specifying different types of text fonts. These tags allow you to enhance the appearance of your Web pages, using bold, italics, and underlining to highlight an item of importance or to illustrate a definition. However, text changes used incorrectly can clutter up a Web page and make it less readable.

7.2.1 Semantic Based Style Types

There are basically two ways to specify how text should be rendered by a browser. (These may be superseded by Cascading Style Sheets in the future.) The first is the *semantic based style type* (also called the *content based style types*), in which HTML tags are used to indicate the content of the text. For example, the text may be an address or a citation, and the browser renders addresses and citations according to built-in settings. You merely tell the browser the category into which the text falls. If you use the address tag, `<ADDRESS>`, for all addresses on a Web page, they will all be rendered similarly. The form of the writing then suggests a meaning to the reader.

A number of the most important semantic based style types, and examples of each, are listed here. Figures 7.1 and 7.2 depict how one browser renders the different examples. On another browser, the items might appear slightly different.

FIGURE 7.1 Samples of Semantic Based Style Types (Part a).

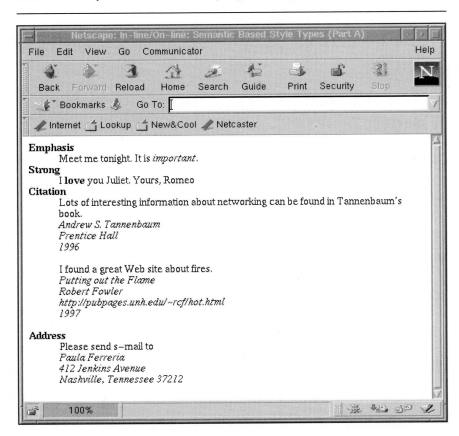

FIGURE 7.2 Samples of Semantic Based Style Types (Part b).

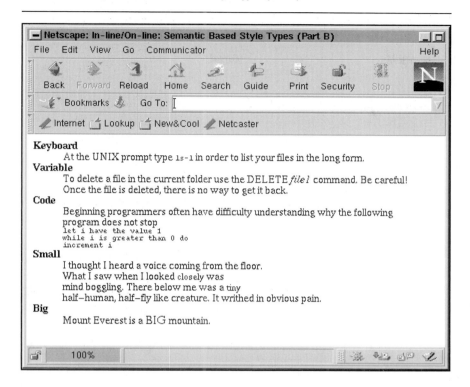

Emphasis Tag

The emphasis tag, `<EM>`, with its corresponding `</EM>` ending tag is used for highlighting text. An example would be:

```
Meet me tonight.  It is <EM>important</EM>.
```

Strong Tag

The strong tag, `<STRONG>`, is used to indicate an even higher level of emphasis. An example is:

```
I <STRONG>love</STRONG> you Juliet.  Yours, Romeo
```

Citation Tag

The citation tag, `<CITE>`, is used to specify a reference. A collection of citations creates a bibliography. Using the citation tag facilitates that collection (possibly automated), since every reference is bracketed between `<CITE>` and

</CITE>. A sample of the citation tag is:

```
Lots of interesting information about networking can
be found in Tannenbaum's book. <BR>
<CITE> Andrew S. Tannenbaum <BR>
Computer Networks <BR>
Prentice Hall <BR>
1996 <BR>
</CITE>
```

Notice that we also use the *line break* tag, <BR>, which forces a line break; that is, the text must go to the next line to continue. There is no ending tag for <BR>.

To cite an on-line reference, people generally include the title, the author, the corresponding URL, and a date. The title is usually extracted from the title bar. Since some Web documents may not include all four of these pieces of information, you should try to include as many as possible. Remember that since Web addresses change frequently, URL citations are not as reliable in the long term as printed matter citations. However, a URL can be more convenient for the reader to track down, because they do not have to go to the library.

Here is an example of citing a Web page.

```
I found a great Web site about fires. <BR>
<CITE> Putting out the Flame <BR>
Robert Fowler <BR>
http://pubpages.unh.edu/~rcf/hot.html <BR>
1997 <BR>
</CITE>
```

A number of documents on the Web suggest citation models. There is no accepted standard, either on-line or in a printed bibliography. The bibliography of this book contains many examples of on-line citations and several more remarks about them.

Address Tag

The address tag, <ADDRESS>, is used to indicate an address. If used throughout a series of Web pages, it is easy to automate the process of developing an address book for the pages. An example of this tag is:

```
Please send s-mail to
<ADDRESS> Paula Ferreria <BR>
412 Jenkins Avenue <BR>
Nashville, Tennessee 37212 <BR>
</ADDRESS>
```

Some authors pretty print their HTML code by including additional spaces. Although this does not affect how the text is rendered by the browser, it could make the source code easier to read, thereby simplifying debugging. For example, the previous address code might be written as:

```
Please send s-mail to
<ADDRESS> Paula Ferreria            <BR>
          412 Jenkins Avenue        <BR>
          Nashville, Tennessee 37212 <BR>
</ADDRESS>
```

Keyboard Tag

This tag is used to delineate keyboard input. For example,

```
At the UNIX prompt type <KBD>ls -l</KBD> in order to
list your files in the long form.
```

It is good practice to use the keyboard tag, <KBD>, for all typed commands. Among other things, this ensures a consistent style for such inputs. Consider the following scenario: You are setting up a group of Web pages that describe how to learn to type more efficiently. Initially, you decide to use bold text for keyboard input, since you really want the keystrokes to stand out. You begin writing your HTML code. A couple days later, you are interrupted by a co-worker, who urgently needs you to complete a different task. When you finally return to your typing page, you have forgotten that you were using bold for keyboard input, and you switch to emphasis. This results in inconsistent styles for keystrokes, which can confuse your readers.

Scenarios such as this are quite common, especially when multiple users are developing a Web presentation together. For this reason, we suggest using the keyboard tag consistently. In general, try to use the semantic based style type that fits the application.

Variable Tag

Computer scientists developed HTML, and many of them like to talk about programming. To do so on Web pages, they introduced the variable tag,

<VAR>, and the code tag, <CODE>, which is discussed in the next section. The variable tag is used to indicate an expression, usually just a sequence of letters, that has a number of different possible values. For example, in the equation

$$x + y = 10$$

x and *y* are variables that can equal 7 and 3, respectively, or 4 and 6, or 1 and 9, etc.

In some operating systems, the DELETE command is used to delete a file. The "argument" to the command is a file name. For example, let the variable name file1 represent any file name. In the on-line documentation we are developing about file manipulation, we can specify how to delete a file, using the following HTML code:

```
To delete a file in the current folder use the
DELETE <VAR>file1</VAR>
command.  Be careful!  Once the file is deleted, there is
no way to get it back.
```

Code Tag

The code tag, <CODE>, is used for specifying program code. For example,

```
Beginning programmers often have difficulty understanding
why the following program does not stop:
      <CODE> let i have the value 1 <BR>
            while i is greater than 0 do <BR>
                  increment i <BR>
      </CODE>
```

Small Tag

Earlier, we saw how to use the basefont and font tags either to increase or decrease the font size. Another method for reducing the *relative font size* is to use the small tag, <SMALL>. Text appearing in this tag is made proportionally smaller than the surrounding text. Here is one possible use:

```
I thought I heard a voice coming from the floor.
What I saw when I looked <SMALL>closely</SMALL> was
mind boggling.  There below me was a <SMALL>tiny</SMALL>
half-human, half-fly like creature.  It writhed in
obvious pain.
```

Do not overuse the small tag, as it can make the text too small to read.

Big Tag

The big tag, <BIG>, plays the opposite role of the small tag.

7.2.2　Syntactic Based Style Types

In contrast to the semantic based style types, *syntactic based style types* (also called *physical based style types*) allow you to tell the browser specifically how you want the text to appear. For example, you may want it to be in bold or italics. When you specify that text is to be in italics, the browser formats that text in italics. Notice the difference between this and an indication that the text should be rendered, say, as an address. How is an address to be formatted? Should it be indented? Bold? Italics? In the semantic based style, the browser takes care of these details. In the syntactic specification, the HTML programmer must describe all of the formatting details.

Here we present the most important syntactic based style types and examples of each. Figure 7.3 depicts how one browser renders the different examples. They might look slightly different on another browser.

Bold Tag

The bold tag, <B>, is used to make text in boldface. Most browsers darken the text and widen the letters. A sample use of the tag is as follows:

```
<B>The Mighty, Mighty Bosstones</B> have some lyrics
that could be considered controversial.
```

In this case, the name of the band stands out on the Web page. The strong tag would be a good choice here as well.

Italics Tag

To place a portion of text in italics, use the italics tag, <I>. For example, italics might be a good choice for the text of a poem:

```
<CENTER>
<STRONG>When I Met You</STRONG> <BR>

<I>
When I met you my life was enhanced. <BR>
When I met you my heart danced. <BR>
When I met you I felt forever blessed. <BR>
Your image came to me as I ran out of time on the test.
</I>
</CENTER>
```

FIGURE 7.3 Samples of Syntactic Based Style Types.

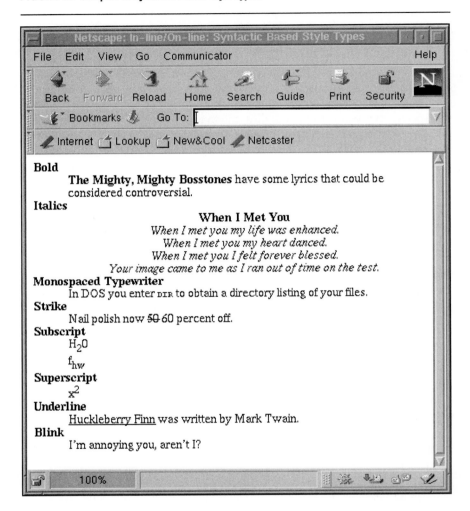

Notice that we also used the center tag, `<CENTER>`, to center the poem horizontally on the screen.

Monospaced Typewriter Text Tag

The typewriter text tag, `<TT>`, is used for placing text in a monospaced typewriter font. This can be used to indicate that a certain phrase needs to be typed in. For example, suppose you are writing a Web page describing how to list your files in DOS. You could do the following:

```
In DOS you enter <TT>DIR</TT> to obtain a directory
listing of your files.
```

Strike Tag

The strike tag, `<STRIKE>`, may be used for crossing out a word by having a line drawn through it. If you have a business Web page and want to indicate that an item is on sale, you can use the strike tag to show that prices have been slashed. For example,

```
Nail polish now <STRIKE>50</STRIKE> 60 percent off.
```

The strike tag is also useful for depicting information that is no longer valid and has been updated. Except in a business situation, like that noted here, and in some specialized legal documents, this tag is rarely used.

Subscript Tag

The subscript tag, `<SUB>`, is used to generate a subscript. Suppose you are producing a Web page for your chemistry class and you want chemical formulas on your Web page to look professional. You might specify the formula for water as follows:

```
H<SUB>2</SUB>O
```

You can also include more than a single character as a subscript. For example, suppose you have a function, f, that depends on both height and weight. You might specify it as:

```
f<SUB>hw</SUB>
```

Superscript Tag

The superscript tag, `<SUP>`, produces on-screen superscripts, working in much the same way as the subscript tag works.

Underline Tag

The underline tag, `<U>`, is used to underline text. Since hyperlinks are depicted by underlining, the underline tag should be used sparingly and only in situations where no confusion can result as to whether or not the underlined item is a hyperlink. Here is a sample use:

```
<U>Huckleberry Finn</U> was written by Mark Twain.
```

Most users tend to try to click on underlined text, even though the underlining may be done in a different color than that of a hyperlink and even if the underlined text is not highlighted, as are hyperlinks.

Blink Tag

Flashing text is created using the blink tag, `<BLINK>`. Because most users find flashing text very annoying, it should be used very sparingly. Here is an example:

```
<BLINK>I'm annoying you, aren't I?</BLINK>
```

7.2.3 Style Type Usage

In theory, you should use semantic based style types wherever possible, because these codes reflect the meaning of the formatted text. For example, if the citation tag is used, it is usually safe to assume that the enclosed text is a reference. It is easy to imagine a computer program that reads an HTML file and outputs a listing of all the citations, creating a bibliography. On the other hand, if the citations are formatted with italics, the computer program would not be able to distinguish the citation from other text also in italics. In practice, most users tend toward using the syntactic based style type, to exercise a higher degree of control over the appearance of their pages.

You may notice a difference in the way various browsers render semantic tags. For example, some browsers may render addresses in italics, while others use bold. For the most part, however, the popular browsers are fairly consistent. If a browser renders an address in italics, why not use the italics tag? Again, at some point, a computer program that uses the address tag to extract an address book automatically could be applied to a Web presentation. Also, it is much easier for the reader if each style element, such as an address, hyperlink, reference, and so on, appears in its own consistent format.

Some older tags you may encounter are called *deprecated tags*. This means that the tag is of little value and will eventually disappear from the HTML standard. Examples of such tags are `<LISTING>` and `<XMP>`. Do not use deprecated tags.

Try not to overuse the various style types. Too many changes can make a document harder to read.

EXERCISES 7.2 Semantic Versus Syntactic Based Style Types

1. Create HTML fragments in which you make effective use of the following tags:

    (a) Emphasis

(b) Strong

(c) Citation

(d) Address

(e) Keyboard

(f) Variable

(g) Code

2. Locate a Web page that is cluttered with too many style tags. Provide its URL and critique it.

3. What is the result of nesting small tags? Experiment and find out if you can make text tiny by nesting several of these. At what point does the text stop getting smaller? Repeat the same experiment for the big tag. Are the results the same?

4. Show how the superscript tag can be used effectively to express numbers in scientific notation.

5. Format your favorite poem using the HTML tags discussed in this section.

6. Supply the URL of a Web page that makes poor use of underlining in the sense that its hyperlinks are hard to distinguish from simple underlined text.

7. Is it possible to underline an item twice? That is, can you produce something like twice underlined?

8. Is it possible to make an entire screen blink?

9. Can you find any more semantic or syntactic based style tags that are supported by your browser? If so, list them and produce a sample use of each.

10. Can you find any other deprecated tags?

7.3   Headers and Footers

7.3.1   Headers

You have probably noticed that many Web pages contain similar types of information at the top. The beginning part of a rendered Web page is called the *header*. The header is the information contained at the top of a rendered Web page, not at the top of an HTML source file. Do not confuse the header with the contents of the head tag. The header is not formatted within the head tag, but in the body of a document. Most headers contain a (nonempty) subset of the following information:

- The title of the page.

- Last-update information.

- Signature of the page developer.

- An icon or logo associated with the page.

- A counter of the number of visitors.

- An advertisement.

The purpose of the header is to convey the most important information about the page, introduce the page, and set the tone for the page. Figure 7.4 shows a screen shot of a sample header. This header consists of a title, an icon, last-update information, and the signature of the page developer, ROXY. In our example, the header information is offset by horizontal lines.

In any collection of related Web pages, it is a good idea to use consistent headers. This helps the reader to determine the boundaries of the presentation. If a hyperlink leads to a different looking header, the reader realizes they may have left the original presentation. Consistent headers help tie the presentation together.

FIGURE 7.4 A Sample Header.

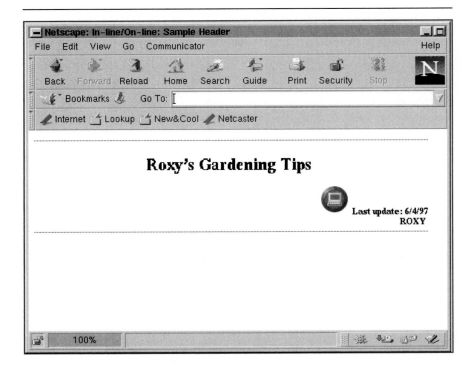

7.3.2 Horizontal Rule Tag

Horizontal lines are produced using the horizontal rule tag, <HR>; there is no accompanying ending tag. To draw a horizontal line all the way across the browser's window, merely include the following HTML code:

```
<HR>
```

The horizontal rule tag has several attributes that are supported by most browsers. They are WIDTH, SIZE, NOSHADE, and ALIGN. The WIDTH attribute is used to specify how wide the horizontal line should be. The default draws a line that crosses the entire width of the browser window, as shown in Figure 7.4. The WIDTH attribute's value can be specified either as a percentage of the browser window's width, or as a fixed number of pixels.[1] Depending on how you are using the horizontal rule tag, one method may be preferable to the other. To draw a horizontal line across 45 percent of the browser's window, use the following HTML code:

```
<HR WIDTH = "45%">
```

To produce a line 150 pixels wide, use:

```
<HR WIDTH = "150">
```

By default, horizontal lines drawn by the horizontal rule tag are about 3 pixels high. The SIZE attribute allows you to define a taller (thicker) line. For example, to draw a horizontal line across 75 percent of the screen that is 10 pixels high, use the following HTML code:

```
<HR WIDTH = "75%" SIZE = "10">
```

The order in which the WIDTH and SIZE attributes appear is not important.

By default, the horizontal rule tag draws what looks like an engraved line. To produce a darker, flatter-looking horizontal line, use the NOSHADE attribute, which functions like a switch and has no value.

When a horizontal line is less than 100 percent of the width of the screen, it needs to be aligned. The ALIGN tag, which accomplishes this function, may have any of the following values: left, right, and center. It is possible to create a tapered-looking graphic by combining all of the attributes of the horizontal rule tag. Figure 7.5 provides an example.

---

[1] Recall that a pixel stands for "picture element" and is a very small unit of measure.

FIGURE 7.5 Illustration of a Tapered-looking Graphic, Created by Using All the Attributes of the Horizontal Rule Tag.

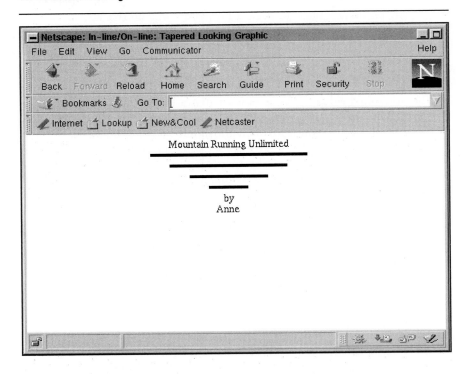

The HTML code used to produce the horizontal lines in Figure 7.5 is as follows:

```
<HR ALIGN = "center" NOSHADE SIZE = "4" WIDTH = "40%">
<HR ALIGN = "center" NOSHADE SIZE = "4" WIDTH = "30%">
<HR ALIGN = "center" NOSHADE SIZE = "4" WIDTH = "20%">
<HR ALIGN = "center" NOSHADE SIZE = "4" WIDTH = "10%">
```

If you decide to include a tapered set of horizontal lines, be careful. The effect you end up with will depend largely on the width of the browser's window.

7.3.3 Footers

The bottoms of many Web pages contain similar types of information. The ending part of a Web page is called the *footer*. Most footers contain a (nonempty) subset of the following information:

- Navigational aids.

- Last-update information.

- The Webmaster's name.

- A `mailto` hyperlink to the Webmaster.

- A hyperlink leading to a *frequently asked questions* (*FAQ*) page.

- A copyright notice.

- A disclaimer.

- A `README` file that usually contains acknowledgements.

- A publication date.

- Advertisements.

The purpose of the footer is to convey additional important information about a page, such as navigational aids, copyright notice, Webmaster, and Webmaster's e-mail address. Figure 7.6 shows a screen shot of a sample footer, which consists of a set of navigational aids, telephone numbers to obtain more information, a couple of hyperlinks for questions, a copyright notice, and a hyperlink providing information about the server used for the presentation. Notice also that the footer is offset by a horizontal line.

Most HTML servers allow the use of keywords to insert a boilerplate in the footer; that is, to insert the date and a hyperlink automatically. However, the conventions for this are server-specific and could affect file portability if someone develops and tests them on one server and then moves them to a different server.

A standard practice for navigation information in a footer is the use of vertical bars or some other symbol to separate the hyperlinks. It is also customary to use a smaller font size. Figure 7.7 shows an example that was rendered from the following HTML code:

```
<H5 ALIGN = "center">
<HR>
<A HREF = "alaska.html">Last Frontier</A> |
<A HREF = "yosemite.html">Yosemite</A> |
<A HREF = "bigsky.html">Glacier Park</A>
</H5>
```

Notice the use of the `<H5>` heading tag, which produces a small font size. We have also used the `ALIGN` attribute of the `<H5>` tag to center the menu

FIGURE 7.6   A Sample Footer.

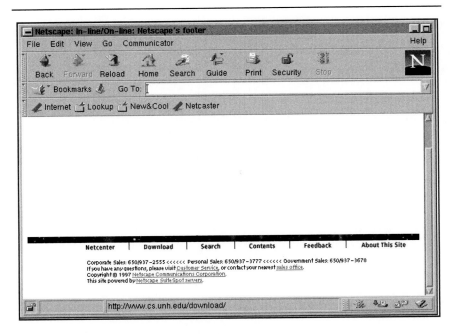

FIGURE 7.7   Hyperlinks Separated by Vertical Bars in a Footer.

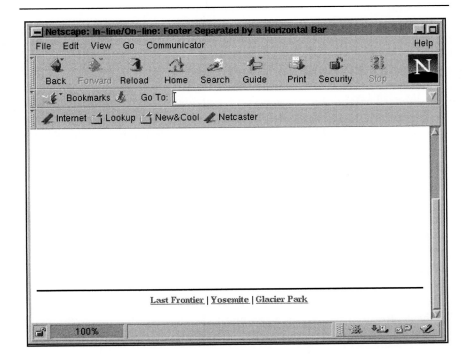

items. The center tag would have done equally well. The division tag, `<DIV>`, with a value of `center` for its `ALIGN` attribute, will also center the hyperlinks. You could also use the `ALIGN` attribute of the paragraph tag with a value of `center`. This type of navigation is very effective and easy to use.

The remarks pertaining to the desirability of consistent headers also apply to footers. The reader then knows what to expect at the bottom of each page. This makes them feel more comfortable with the presentation.

EXERCISES 7.3 Headers and Footers

11. Design a sample header suitable for your Web pages.

12. Design a tapered-looking set of five horizontal lines to go under the phrase "These are the days to remember."

13. Can you "black out" an entire screen by drawing a wide enough horizontal line and setting the appropriate attributes?

14. Design a sample footer suitable for your Web pages.

7.4   Lists

Three primary types of lists are available in HTML: *ordered lists*, *unordered lists*, and *definition lists*. Each list environment is well-suited for one or more specific types of writing. The entire list is surrounded by a beginning and an ending list tag, which varies depending on the type of list. In addition, list item tags are used to identify each entry in the list. The list item tag also depends on the type of list you are using.

7.4.1   Ordered Lists

In an *ordered list*, the elements are prefixed by a symbol that denotes their relative order within the list. The most commonly used symbols for marking the elements of an ordered list are Arabic numbers, letters, and Roman numerals. An ordered list is used for a series of sequential steps or specifically ordered items. The beginning and ending tags for an ordered list are `<OL>` and `</OL>`, respectively. The beginning tag for each list item is `<LI>`. No ending tag is necessary for an item, since the browser can determine the end of a list item by encountering either another list item tag, `<LI>`, or an ending tag for the list, `</OL>`. By default, the items in an ordered list are numbered using Arabic numbers. The following HTML code illustrates the use of an ordered list to describe the process of walking a dog.

```
<H3>How to Walk a Dog</H3>

<OL>
    <LI> Call the dog.
```

```
    <LI> Call the dog with a dog bone in hand.
    <LI> Feed the dog and attach the leash to the dog's
         collar.
    <LI> Walk, stop, pull, walk, stop, pull.
    <LI> Return home.
    <LI> Remove the leash from the dog's collar.
    <LI> Give the dog a biscuit and pat it on the head.
</OL>
```

Note that the order of the elements in this list is important. Thus, the ordered list environment is necessary. Figure 7.8 shows how this code is rendered by a browser.

What if you want to use characters or Roman numerals instead of Arabic numbers to label the items? This can be done by using the TYPE attribute of the ordered list tag. TYPE can take the values a, A, i, and I. The lowercase a results in labels a, b, c, and so on. Similarly, the uppercase A results in labels A, B, C, and so forth. The i or I TYPE attribute values generate either lowercase or uppercase Roman numerals, respectively. For example, i results in a numbering of i, ii, iii, iv, v, and so on.

It is also possible to begin numbering from a value other than 1. The starting value can be defined with the START attribute of the list item tag. For example, to generate Arabic numbers starting from a value of 4 instead of 1,

FIGURE 7.8 An Example of an Ordered List That Describes the Process of Walking a Dog.

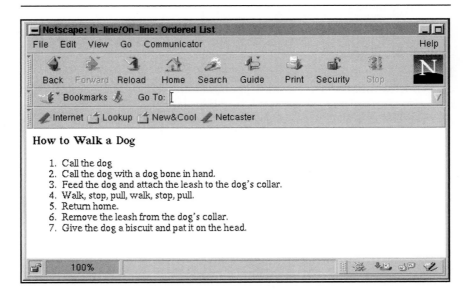

you would use the following code:

```
<OL START = "4">
    <LI> Step four.
        ...
</OL>
```

Many books use this type of numbering in the exercises. If you want to use lowercase characters starting from e, you would use the following code:

```
<OL TYPE = "a" START = "5">
    <LI> Step e.
        ...
</OL>
```

The starting values for Roman numerals can be similarly adjusted.

The list item tag also has an attribute. The VALUE attribute of the list item tag can be used to change an item's label. To generate an ordered list whose items are numbered 1, 2, 5, and 6, the following code would be used:

```
<OL>
    <LI> Step 1.
    <LI> Step 2.
    <LI VALUE = "5"> Step 5.
    <LI> Step 6.
</OL>
```

Note that once you change an item's value, the succeeding items continue on from that new value. The VALUE attribute may be used in a similar fashion to change the "numbering" when letters or Roman numerals are used.

7.4.2 Unordered Lists

The ordered list is used when the order in which the items appear is significant. If the order is not important, you can use the *unordered list*. Rather than being numbered, the items are usually marked with *bullets*.

The beginning and ending tags for an unordered list are <UL> and </UL>, respectively. Each item in an unordered list is identified by a list item tag, as it was for an ordered list. The following HTML code describes some "House Sitting Chores" using an unordered list:

```
<H3>House Sitting Chores</H3>

<UL>
      <LI> Pick up mail.
      <LI> Walk the dog.
      <LI> Water geraniums.
      <LI> Feed the cat.
</UL>
```

Figure 7.9 shows how a browser renders this code. Notice that the items are not interdependent; that is, they can appear in any order.

Each item is prefaced by a bullet, and the default bullet is a shaded disk. It is also possible to use other bullets in HTML. You can use the TYPE attribute of the unordered list tag to generate an open circle or an open square. For example, the following HTML code would use open circles as bullets:

```
<H3>House Sitting Chores</H3>

<UL TYPE = "circle">
      <LI> Pick up mail.
```

FIGURE 7.9 An Example of an Unordered List That Describes Some House Sitting Chores.

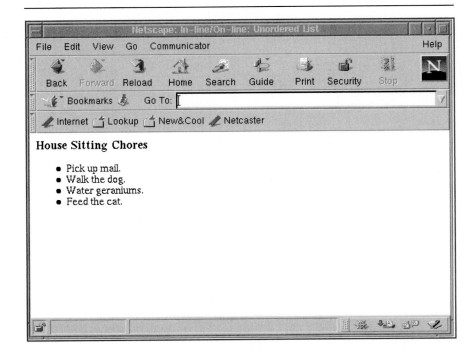

```
    <LI> Walk the dog.
    <LI> Water geraniums.
    <LI> Feed the cat.
</UL>
```

You can also include "custom" bullets, which is best done using *definition list*, as discussed in the next section.

7.4.3 Definition Lists

Definition lists are unordered lists in which each item has two parts: a *term*, and its corresponding definition. The beginning and ending tags for the definition list are <DL> and </DL>, respectively.

Instead of using a list item tag for each item, the term portion requires a definition term tag, <DT>. The definition portion uses a data definition tag, <DD>. The corresponding ending tags for the definition term and data definition tags are </DT> and </DD>, respectively. However, they are rarely used.

The following HTML code provides an example of a definition list used to describe running races:

```
<H3>Common Foot Race Distances</H3>
<DL>
    <DT> <STRONG>5K</STRONG>
    <DD>A sprint.

    <DT> <STRONG>10K</STRONG>
    <DD>A quick, hard run.

    <DT> <STRONG>Marathon</STRONG>
    <DD>A 26.2 mile run (hurts over the last 6 miles).

    <DT> <STRONG>50K</STRONG>
    <DD>A short Ultra of 31 miles.

    <DT> <STRONG>Century</STRONG>
    <DD>A 100 mile run.
</DL>
```

Figure 7.10 shows how the code is rendered. Notice that the browser placed each term on its own line, and the definition is on the succeeding line and is slightly indented. Also observe that we used the strong tags for the terms to

FIGURE 7.10 An Example of a Definition List Used to Describe Common Running Races.

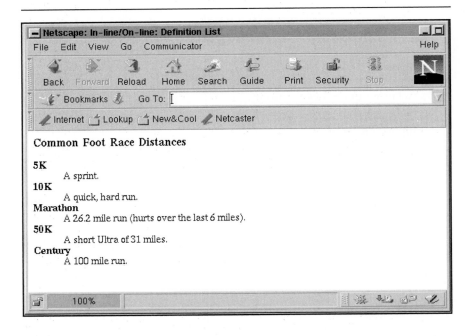

be defined: 5K, 10K, Marathon, 50K, and Century. This makes them stand out better.

The definition list environment is often used for displaying a glossary, or for defining a group of terms. Another common use of the definition list tag is the presentation of a group of items with custom bullets. Suppose you want to produce a page, located in your `public-html` directory, that contains a list of your textbooks for the semester. You would like to include a custom bullet called `purplepin.gif`, which sits in your `GIF` directory, which in turn is located in your `public-html` directory. The following HTML code does what you want:

```
<H3>My Textbooks This Semester</H3>

<DL>
    <DT> <IMG SRC = "../GIF/purplepin.gif"
            HEIGHT = "21"
            WIDTH = "21"
            ALT = "*">
    <DD>Deutsch Heute
```

```
<DT> <IMG SRC = "../GIF/purplepin.gif"
            HEIGHT = "21"
            WIDTH = "21"
            ALT = "*">
<DD>In-line/On-line

<DT> <IMG SRC = "../GIF/purplepin.gif"
            HEIGHT = "21"
            WIDTH = "21"
            ALT = "*">
<DD>U. S. History

<DT> <IMG SRC = "../GIF/purplepin.gif"
            HEIGHT = "21"
            WIDTH = "21"
            ALT = "*">
<DD>The Complete Works of Shakespeare
</DL>
```

Figure 7.11 shows how this code is rendered.

Some Web authors omit the data definition tag and simply place the definition on the same line as the definition term tag.

7.4.4   Nested Lists

It is possible to *nest* the different types of lists within one another, or to nest multiple lists of the same type. In computer science, the word "nest" is used to indicate layers within layers. Think of an onion; when the outer layer is removed, a new inner layer is exposed. The inner layer is completely nested inside the outer layer. When nesting list environments (or any environments, for that matter), you should not let them overlap. An example of a poorly written nesting sequence is as follows:

```
<H3>How to Make a Tossed Salad</H3>

<OL>
    <LI> Open Martha Stewart book on the subject.
    <LI> Wash veggies.
    <LI> Dry veggies.
    <LI> Slice veggies.
    <UL>
```

FIGURE 7.11 An Example of a Definition List That Uses Custom Bullets.

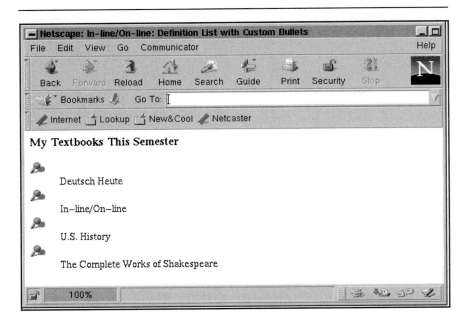

```
      <LI> Cut tomatoes into wedges.
      <LI> Tear lettuce into small pieces.
      <LI> Thinly slice the radishes.
      <LI> Grate carrots.
      <LI> Slice cucumbers.
   </OL> <!-- This is incorrectly nested. -->
</UL>
```

The inner list environment must be closed before the outer list environment is closed. In our incorrectly coded example, the opening unordered list tag is mismatched with the closing ordered list tag, and the opening ordered list tag is misaligned with the closing unordered list tag. When nesting different types of lists, make sure you close them off properly.

Nested lists are useful when you would like to expand on one or more items. The following code illustrates nested ordered and unordered lists.

```
<H3>UNH Scores Its First Goal at A Hockey Game</H3>

<OL>
   <LI> Puck goes past opposing goalie.
   <LI> Goalie yells.
```

```
    <LI> Sticks are raised.
    <LI> Puck enters net.
    <LI> Red light comes on.
    <LI> Referee heads for net.
    <UL>
        <LI> Someone grabs for a
            <EM>long, frozen fish</EM>.
        <OL TYPE = "a">
            <LI> Fish is thrown.
            <LI> Fish clears boards.
            <LI> Fish hits ice.
            <LI> Doors at side of rink open.
            <LI> Man with bucket appears.
            <LI> Fish is scooped up into bucket.
            <LI> Man and fish leave.
            <LI> Doors at side of rink close.
            <LI> Goalie watches.
        </OL>
        <LI> Fans go wild.
        <OL TYPE = "a">
            <LI> Students go berserk.
            <LI> "Sieve" (to make the opposing goalie
                    feel at home) is shouted.
            <LI> Fingers are pointed indicating number
                    of goals scored, here "1."
            <LI> Goalie slaps his pads with stick.
            <LI> Goalie pretends to drink water.
        </OL>
        <LI> Referees showboat.
        <OL TYPE = "a">
            <LI> Referees skate wildly around the ice.
            <LI> Referees fix their hair.
        </OL>
    </UL>
    <LI> Announcement of goal scorer's name.
    <LI> Fans remain standing.
    <LI> New face-off.
</OL>
```

Figure 7.12 shows how this code is rendered.

FIGURE 7.12 An Example of Nested Ordered Lists.

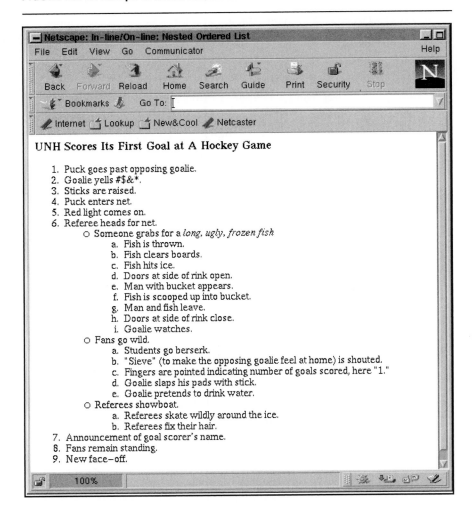

At most levels, ordered lists are used because the steps actually happen in a certain sequence. We use the TYPE attribute of the ordered list tag to change the inner layers from Arabic numbering to letters. The first nested list is unordered. It indicates that the three items involving fish, fans, and referees all happen in the same timeframe. Each of these is further divided into a sequence of steps. Notice how indentation is used to make it easy to line up the beginning and ending list tags.

Not all nested lists are as elaborate as the example shown in Figure 7.12. In fact, it is usually a good idea not to nest lists more than one or two levels deep. Also, it is common practice to use a mix of Arabic numbers and letters. Another popular setup is uppercase letters for the outer level and lowercase letters for the nested items. Alternatively, you could use capital Roman numerals for the outer level and lowercase Roman numerals for the inner level. It is generally not a good idea to blend Arabic numbers, letters, and Roman numerals. The three combined can create an unappealing and hard to follow list.

If the items in a group of nested lists do not have any inherent order, there is no reason to assign one artificially. In this situation, you may want to use unordered lists. The following HTML code is an example:

```
<H3>Critters</H3>

<UL>
    <LI> <STRONG>Mammals</STRONG>.
    <UL>
        <LI> Blue whale.
        <LI> Cat.
        <LI> Water buffalo.
    </UL>
    <LI> <STRONG>Reptiles</STRONG>.
    <UL>
        <LI> Alligator.
        <LI> Lizard.
        <LI> Snakes.
        <UL>
            <LI> Boa.
            <LI> Cobra.
            <LI> Python.
        </UL>
    </UL>
    <LI> <STRONG>Insects</STRONG>.
    <UL>
        <LI> Dobb's Fly.
        <LI> Hornet.
        <LI> Mosquito.
    </UL>
</UL>
```

Figure 7.13 shows how a browser displays this code. Notice that, by default, a shaded disk is used at the outer layer, an open circle is used for the middle items, and an open square is used at the innermost layer. These symbols can be changed using the TYPE attribute of the unordered list tag.

FIGURE 7.13 Example of Nested Unordered Lists.

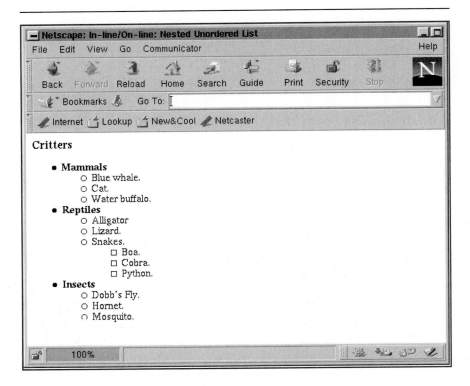

Definition lists, as well as any combination of definition, ordered, and un-ordered lists, can all be nested. It is uncommon though to see all three mixed together, as this can result in a cluttered and "over-organized" look.

EXERCISES 7.4 Lists

15. When would it be useful to begin an ordered list at a value other than 1? Give an example.

16. Create an ordered list of the five lightest chemical elements.

17. Code an ordered list that uses lowercase Roman numerals, starts at a value of 9, and contains four elements.

18. Create an unordered list containing hyperlinks to five "cool" Web pages.

19. Using custom bullets, code a definition list which describes your favorite four dinner entrées, as well as the best local restaurants where they can be found.

20. Find five interesting pieces of Internet jargon on the Web. Create a definition list with custom bullets, in which the terms and their definitions are given.

21. Present an application of nested ordered lists and code it.

22. What happens if you nest an unordered list more than three levels deep? Are some of the bullet types reused?

23. Present an application of nested unordered lists and code it.

24. Are there applications where it makes sense to use both ordered and un-ordered lists together? If so, describe and code one.

7.5 Tables

Tables in HTML pages allow you to organize information in a row and column format. For example, you might want to display your courses and their corresponding grades in a tabular form on your resume. Table 7.1 illustrates one possible layout of such information. The horizontal and vertical lines are called (table) *dividers* (see Figure 7.15). Using HTML tables, you could display this information in a similar fashion in your on-line resume.

In addition to laying out information in a tabular format, HTML tables are used to present any type of information for which you want a lot of control over the positioning of the material. For example, HTML tables could be used to achieve a newspaper-like appearance. We usually do not think of a newspaper as being in a tabular format, but a newspaper format can most easily be produced on a Web page using tables with the dividers suppressed. Tables have many uses, some nonintuitive, for achieving specific layouts.

7.5.1 Table Usage

There are many ways to use tables to format information on the Web. When you see an interesting layout on a Web page, view the source code and see how the effect was achieved. There is a good chance it was produced using tables. The following are some situations for which you might use tables:

TABLE 7.1 A Sample Tabular Format Consisting of a Student's Courses and Grades.

Fall Semester		Spring Semester	
Business	B−	Accounting	B−
German 1	C	German 2	C+
Math	A	In-line/On-line	A
Sociology	B+	Philosophy	A−
Tennis	A+		

- *Present tabular information*—If you have information or data that is naturally divided into rows and columns, it can probably be easily and effectively formatted using HTML tables.

- *Control layout*—If you want to control the layout of text, position a group of images, or present an extensive menu, you may decide to use tables to achieve the desired appearance.

- *Express relationships*—If you need to display relationships between a group of items, tables are usually a good mechanism to use. For example, in Table 7.2, we show whether a number of people can play the guitar or piano, and whether they have acting experience. Using the table, it is easy to locate someone, namely Jan, who plays both the guitar and the piano, and also has some acting experience. If a table is laid out properly, relationships between elements can be expressed clearly, and "new" relationships can be easily discovered.

(In Figures 7.20 and 7.21, we show how Tables 7.1 and 7.2 appear when formatted in HTML).

7.5.2 HTML Table Tags

Rows and Columns

Tables are created using the table tag, `<TABLE>`, with the ending tag `</TABLE>`. The most important tag that goes inside the table tag is the table row tag, `<TR>`. Its corresponding ending tag, `</TR>`, is often omitted. Browsers can

TABLE 7.2 A Table Illustrating the Relationships Between People, Musical Instruments, and Acting Experience.

Name	Play Guitar	Play Piano	Acting Experience
Alice	YES	NO	NO
Claude	NO	NO	YES
Heidi	NO	NO	NO
Jan	YES	YES	YES
Jean-Yves	NO	NO	NO
Mara	YES	YES	NO
Nadine	YES	NO	NO
Pearl	NO	YES	YES
Sun-Lee	NO	YES	YES
Woo	NO	NO	YES

determine where the next row starts by encountering another beginning table row tag. When the browser sees the </TABLE> tag, the table is ended.

The items in a row are specified using the table data tag, <TD>. Its corresponding ending tag, </TD>, may be omitted, for the same reason that </TR> can. We sometimes refer to the location where the elements in table data tags appear as *table cells* (see Figure 7.18). Nearly any HTML element can appear in a table data tag.

There is no table column or <TC> tag. Instead, the number of columns is determined by the row containing the most items, expressed by table data and table header tags (discussed later in this chapter). If any row has fewer than the maximum number of items, its elements will be positioned starting from the leftmost column and continuing to the right, as necessary.

The final version of Table 7.1 requires some experience to produce in HTML. Therefore, we will generate this table incrementally by adding features to a simple starting version. Our goal is not to produce an exact copy of Table 7.1, but rather incrementally develop a table that contains the same information and looks good on-line.

Figure 7.14 shows a screen shot of a five-row, four-column version of Table 7.1. (We did not count the boldface headings as a row.)

This table was produced by the following HTML code:

```
<CENTER>
<TABLE>
    <TR>
        <TD> Business
        <TD> B-
        <TD> Accounting
        <TD> B-
    <TR>
        <TD> German 1
        <TD> C
        <TD> German 2
        <TD> C+
    <TR>
        <TD> Math
        <TD> A
        <TD> In-line/On-line
        <TD> A
    <TR>
        <TD> Sociology
```

FIGURE 7.14 A Simple Version of Table 7.1.

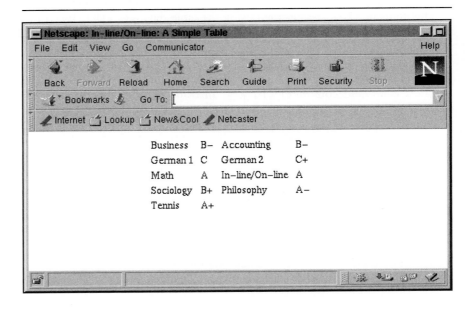

```
            <TD> B+
            <TD> Philosophy
            <TD> A-

        <TR>
            <TD> Tennis
            <TD> A+
    </TABLE>
    </CENTER>
```

Notice that we used the center tag to center the table on the page. You will probably want to center most tables, especially if you decide not to let text flow next to them. Most browsers do not support a value of center for the ALIGN attribute of the table tag. You should therefore use either <CENTER>, <P> with a value of center for the ALIGN attribute, or <DIV> with a value of center for the ALIGN attribute.

The ALIGN attribute of the table tag with a value of left will position the table at the left edge of the page; this is the default positioning. If text appears adjacent to the table declaration, it will flow along the right edge of the table.

With a value of `right` for the ALIGN attribute, the table will be positioned flush right and text will flow along the left side of the table.

In many situations, the text crowds the table, since the browser leaves very little space between the table and the text. It is therefore often desirable to add some space between the text and the table. To add horizontal space, you can use the HSPACE attribute of the table tag, with a pixel value indicating how much space to add. The VSPACE attribute works similarly, controlling the amount of vertical space between the table top and bottom and the adjacent text.

In contrast to Tables 7.1 and 7.2, the table shown in Figure 7.14 does not have a border. You can easily add a border using the BORDER attribute of the table tag. When used without an argument, the BORDER attribute adds a 2-pixel-wide perimeter and 2-pixel-wide row and column dividers. For example, the following code creates a table partitioned by 2-pixel-wide lines:

```
<TABLE BORDER>
```

Including a border makes it easier to see the row and column divisions. The table cells appear as enclosed rectangles. The BORDER attribute can have a pixel value as an argument. The following code adds a border 3 pixels wide around the table:

```
<TABLE BORDER = "3">
```

This is added to the 2-pixel-wide dividers between each consecutive pair of rows and each adjacent pair of columns. The result is shown in Figure 7.15.

Notice how the second half of the fifth row is blank. If you would like to include the partitions here, you can add two *placeholder* table data items. This is usually done as follows:

```
<TR>
    <TD> Tennis
    <TD> A+
    <TD>  
    <TD>  
```

The expression   is the *character entity* for a nonbreaking space.

Element Spacing

The information in the table shown in Figure 7.15 looks crowded. The CELLSPACING and CELLPADDING attributes of the table tag allow you to control

FIGURE 7.15 The Table from Figure 7.14, Shown with a 3-Pixel-Wide Border.

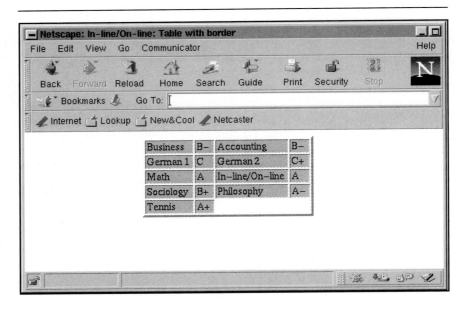

the width of the dividers between rows and columns and the amount of space between the information in a cell and the borders of the cell. The following code has the effect of "spreading out" the information in the table, because the table dividers are set to be 6 pixels wide:

```
<TABLE BORDER = "3" CELLSPACING = "6">
```

The table's perimeter is also made 6 pixels thick. Figure 7.16 shows the effects of these changes. The information is more readable, but it is still not laid out well.

What the table in Figure 7.16 really needs is additional space between each element in the table and the dividers surrounding it. For example, the item "Business" is too close to the table dividers. The CELLPADDING attribute expands the space or "padding" around items in each cell of the table. The effect of the following code on the table is shown in Figure 7.17:

```
<TABLE BORDER = "3" CELLPADDING = "12">
```

The widest item in a column determines the width of that column. A CELLPADDING attribute with a value of 12 has the effect of adding 12 pixels all around the widest element in a column. For example, the course entitled

FIGURE 7.16 **The Table from Figure 7.14, Shown with a 3-Pixel-Wide Border and a** CELLSPACING **Attribute Value of** 6.

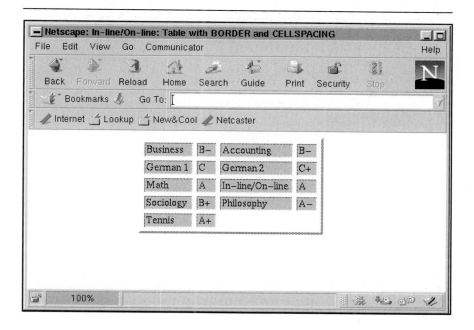

"In-line/On-line" is the widest item in column 3. Twelve pixels of space are added around "In-line/On-line." Figure 7.18 illustrates the effect. The result is that the column appears as though more than 12 pixels have been added around the shorter items.

Figure 7.18 also shows what parts of a table are affected by the CELLSPACING and CELLPADDING attributes. To recap, CELLSPACING affects the table dividers, and CELLPADDING affects the size of the table cells.

The items in the table shown in Figure 7.17 now appear spaced out a bit too far. By combining CELLSPACING and CELLPADDING with the values shown in the following, we obtain a table that is more visually appealing. The declaration

```
<TABLE BORDER = "3"
       CELLSPACING = "6"
       CELLPADDING = "12">
```

can be used to combine the attributes. The order in which these attributes appear is not significant. However, it is a good idea to use a consistent style. The resulting table is shown in Figure 7.19.

FIGURE 7.17 **The Table from Figure 7.14, Shown with a 3-Pixel-Wide Border and a** `CELLPADDING` **Attribute Value of** `12`**.**

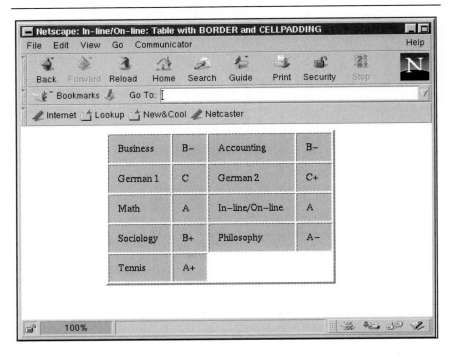

FIGURE 7.18 **The Effect of a** `CELLPADDING` **Attribute of** `12` **on the Widest Cell of a Column.**

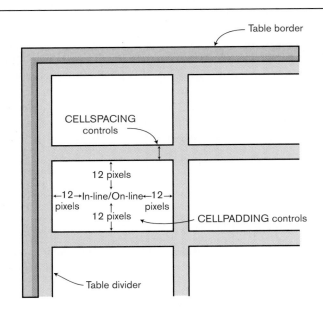

FIGURE 7.19 The Table from Figure 7.14, Shown with a 3-Pixel-Wide Border, a CELLSPACING **Attribute Value of** 6, **and a** CELLPADDING **attribute value of** 12.

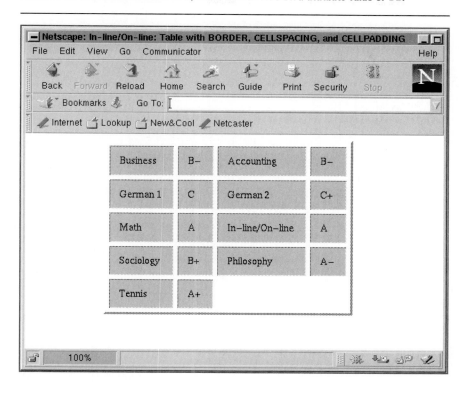

Table Headers

The columns in Table 7.1 have the headers, **Fall Semester** and **Spring Semester**. Notice that the headers are boldface and centered. It is often desirable for tables to include headers and to format them this way. The table header tag, ⟨TH⟩, is useful for this. The following code shows how to add headers to the table we have been constructing:

```
<TABLE BORDER = "3"
       CELLSPACING = "6"
       CELLPADDING = "12">
    <TR>
        <TH COLSPAN = "2"> Fall Semester
        <TH COLSPAN = "2"> Spring Semester
        ...
</TABLE>
```

FIGURE 7.20 The Final Version of Table 7.1 Rendered in HTML.

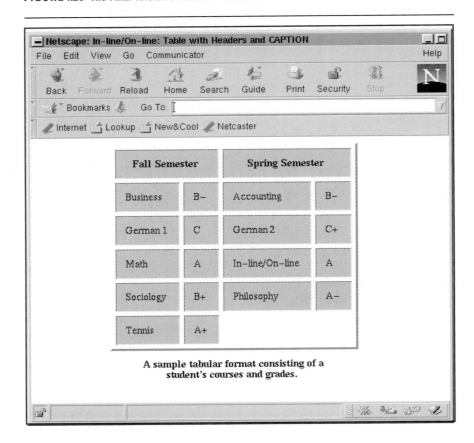

An inspection of Table 7.1 reveals that the headings each span two columns. To achieve the same effect in HTML, we use the COLSPAN attribute of the table header tag. The value of this attribute is a number indicating how many columns the header is to span. The result of this code, which is our final version of Table 7.1, is rendered in Figure 7.20.

There is also a ROWSPAN attribute to the table header tag, and it behaves as you would expect. In fact, COLSPAN and ROWSPAN can be used in a manner similar to attributes of the table data tag. For example,

```
<TD ROWSPAN = "2">
```

specifies that a table item spans two rows. In general, the table data and table header tags possess the same attributes.

Item Positioning

In Table 7.2, notice that in columns 2 through 4, the information is centered, but in column 1, the information is flush left. By default, information included in a table data tag is set flush left, whereas information included in a table header tag is centered. Both of these tags possess the ALIGN attribute, which is useful for positioning material in other ways. The values for ALIGN are left, center, and right. The table row tag also supports the ALIGN attribute and the same set of values. It can be used to position all items in a row at one time.

The code that follows shows one way that the first three rows of Table 7.2 could be produced in HTML. There are many other ways of producing them. For illustration purposes, we have coded each row differently, although we would normally never code using such an inconsistent approach.

```
<!-- do not code using an inconsistent style -->
<TABLE BORDER>
    <TR ALIGN = "center">
        <TD> Name
        <TD> Play Guitar
        <TD> Play Piano
        <TD> Acting

    <TR ALIGN = "left">
        <TD> Alice
        <TD ALIGN = "center"> YES
        <TD ALIGN = "center"> NO
        <TD ALIGN = "center"> NO

    <TR ALIGN = "center">
        <TD ALIGN = "left"> Claude
        <TD> NO
        <TD> NO
        <TD> YES

    ...
</TABLE>
```

Since the table headings were not in boldface in Table 7.2, we decided to produce them using the table row tag instead of the table header tag. It is important to note that the ALIGN attributes of the table data and table header tags override the ALIGN attribute of the table row tag.

In the first row, each item is centered, so we used the center value for the ALIGN attribute of the table row tag. This centered all four table data elements in row 1. In the second row, we used left as the value for the ALIGN attribute of the table row tag. However, because the last three columns have centered items, we had to use the ALIGN attribute of the table data tag, with a value of center, to center them. For the third row, we used an ALIGN attribute value of center for the table row tag and therefore only needed to use a single ALIGN attribute to the table data tag. We did this so we could flush left the item in the first column. Rows 1 and 3 are coded cleanly, whereas row 2 is not. Figure 7.21 shows how Table 7.2 appears when rendered.

FIGURE 7.21   The Final Version of Table 7.2 Rendered in HTML.

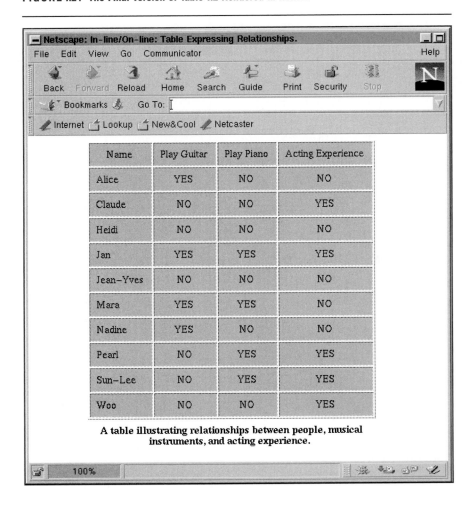

Name	Play Guitar	Play Piano	Acting Experience
Alice	YES	NO	NO
Claude	NO	NO	YES
Heidi	NO	NO	NO
Jan	YES	YES	YES
Jean–Yves	NO	NO	NO
Mara	YES	YES	NO
Nadine	YES	NO	NO
Pearl	NO	YES	YES
Sun–Lee	NO	YES	YES
Woo	NO	NO	YES

A table illustrating relationships between people, musical instruments, and acting experience.

Table Captions

It is often important to annotate a table by adding a caption. For example, all tables shown in this book have captions. You can also create captions for HTML tables. Captions are produced using the beginning and ending caption tags, `<CAPTION>` and `</CAPTION>`. The default setting positions the caption at the top of the table. However, the `ALIGN` attribute of the caption tag can be used to position the caption at the bottom of the table, by using a value of `bottom`. Specifying a value of `top` for the `ALIGN` attribute will result in the default setting.

The caption tag is best placed either immediately after the table tag or immediately inside the ending table tag. Whichever approach you adopt, use it consistently. The following code places the caption directly underneath the table:

```
<TABLE BORDER = "3"
       CELLSPACING = "6"
       CELLPADDING = "12">

<CAPTION ALIGN = "bottom">
    <STRONG>
        A sample tabular format consisting of
        a student's courses and grades.
    </STRONG>
</CAPTION>
    ...
</TABLE>
```

Notice that we have formatted the caption using the strong tag in order to make the caption stand out.

We should mention that most browsers center captions automatically. With a long caption and a narrow table, this usually creates several lines of ragged-left, ragged-right, centered text. If you have a fixed-width table, you may try using the paragraph tag and its `ALIGN` attribute to reposition the text so that it looks better (i.e., justified on one side). We recommend keeping captions short in on-line tables.

Table Width

The table tag has a `WIDTH` attribute that allows you to control the width of the table measured from the outer perimeter, including borders. The attribute takes either a pixel value or a percentage value. For example, the following code instructs the browser to produce a table 600 pixels wide:

```
<TABLE WIDTH = "600">
```

If the table is actually more than 600 pixels wide, the browser ignores the setting. If the browser's screen is less than 600 pixels wide, the reader will need to scroll to see the entire table.

The following code instructs the browser to produce a table that takes up 60 percent of the browser's window:

```
<TABLE WIDTH = "60%">
```

If the table is actually larger than 60 percent of the screen, the WIDTH specification has no effect. Using a percentage setting means that the table size will scale according to the browser's window size. A value of 100% means that the table will scale to occupy the entire width of the browser window, unless it is already larger than this.

Many Web designers specify a pixel value for the WIDTH attribute so that they can obtain more control over their table layouts. If a user has a narrow browser window, this can force the reader to have to scroll to see the entire table. However, if a table is rendered too narrow, the material intended to fit in a single row of a table cell may end up taking more than one "row." This can lead to an undesirable overall appearance. When you see a table on a Web page, experiment with different sized browser windows and observe the effect on the table. The goal is to make the table look good regardless of the screen size.

Table Column Widths

Both the table data and table header tags have a WIDTH attribute. Using this attribute, you can create equal-sized or proportionately sized column widths.

As with the table tag's WIDTH attribute, the WIDTH attribute of the table data and table header tags can take either a pixel value or a percentage value. The WIDTH attribute is used most effectively when the items in the column are not really wide and you would like to scale up the widths of the columns to achieve a more uniform appearance. For example, the following code produces a table in which the first column width is 20 percent of the screen, the second column width is 40 percent of the screen, and the third column width is 20 percent of the screen:

```
<TR>
    <TD WIDTH = "20%"> Butter
    <TD WIDTH = "40%"> English Muffin
    <TD WIDTH = "20%"> Jelly
```

This assumes that:

1. None of the items in the corresponding columns is wider than these percentages.

2. These are the only WIDTH attributes specified for the table data or table header tags.

3. No other row contains more than three items.

Notice in the specification that the value 40 was chosen because

$$40 = 2 \times 20$$

which helps create a more uniform appearance than would a value of, say, 47 for the width of the middle column.

It is good programming practice to include the WIDTH attributes in the table header tags or in the first table data tags, rather than to bury these settings somewhere else in the table. It is also a good idea to specify these values only once for each column. If you do specify more than one setting, or if the items in a column are actually wider than the setting specified, the widest setting or actual size is used, whichever is larger.

If you need to control the exact width of a column, you should use a pixel value, not a percentage. You may want to do this when you are including an image in the table cell or when you want more control of the text layout.

Vertical Alignment

On occasion, you may generate a table with a "tall" row(s). The VALIGN attribute of the table row, table data, and table header tags allows you to position the text vertically within the row. The default positioning centers the material vertically in a row. The most common values for VALIGN are top, center, and bottom. They have the expected positioning effect. The scope of the VALIGN attribute is just the row in which it occurs.

As usual, if a VALIGN attribute is specified for a table row tag <TR> and also for a table data tag <TD> in the same row, the VALIGN attribute of the table data tag will override that of the table row tag.

It is generally a good idea to position all text in a given row at the same level, rather than in a step-wise pattern; even text across a table is easier to read and in most cases is more visually appealing.

Colors

Many browsers support the use of colors in tables. The BGCOLOR attribute of the table, table data, table header, and table row tags can be used to add color to any or all parts of your table. Again, however, settings at the table level can be overridden by those at the table row level, which in turn can be overridden by those at either the table data or table header levels. The allowable values for the BGCOLOR attribute are the same as those for the BGCOLOR attribute of the body tag. For example, the following code produces a white table:

```
<TABLE BGCOLOR = "#FFFFFF">
```

For most tables, it is a good idea to restrict yourself to one or two colors, and those colors should go well with each other and with your page's background pattern or color. In rare situations, more than two colors can be used effectively. For example, in a borderless table, if each entry is colored differently, a patterned effect can be created in which each item stands out. However, the colors must be carefully chosen. This style can be especially effectively when each table data item is a hyperlink.

The on-line presentation accompanying this book compares many effective and ineffective uses of color in tables, as well as a number of more complex table formats.

HTML Table Evolution

You should be aware that not all browsers support all table features described in this chapter. In addition, some browsers support new features not mentioned here. To be completely current with what is available for tables, look on the Web. The standards change quickly. The on-line Web presentation accompanying this book provides information on where to find the most recent data.

7.5.3   Frequently Asked Questions

*   *How can I produce a newspaper-style layout using tables?*

    The key idea is to use a single row with very long columns. So, for a three column newspaper, you would use three table data tags in the first row. The text for the first column follows the first table data tag; the text for the second column follows the second table data tag; and the text for the third column follows the third table data tag. If you want to achieve balanced column lengths, you will need to adjust the amount of text following each table data tag.

    The code described will produce three separate columns. However, you will probably want to include the `VALIGN` attribute of the table row tag, with a value of `top`, so that all text starts at the top of each column. In addition, you may want to add a couple of equal-sized placeholder columns for spacers between the text. This can be done using the following code:

    ```
    <TD WIDTH = "5%">
    ```

    Finally, you will want to set the `BGCOLOR` attribute of the table tag to `#FFFFFF` (white) and the text color to `#000000` (black).

*   *One of the items in my column is making the whole column too wide. What can I do to solve this problem?*

    You can insert a line break tag somewhere in the long item, which will force the rest of the text in that item to move down to the next line. The text will stay in the same cell; it will not go into the next cell down. In some situations, other options might be to split the item over two columns, or use a shorter phrase.

- *Can I nest tables?*

    Yes, and nesting tables is straightforward. In the most common form of nesting, the table tag is included inside a table data item. So, if you want, each table cell can itself be a table.

- *Can I include images and hyperlinks in tables?*

    Yes, you can include them in any table cell, using the tags you would normally use for including these features in an HTML document.

- *Are HTML editors useful for producing tables?*

    Yes, they can be helpful, particularly if you are producing a very simple and regular table. By this we mean the table has the form of a grid. However, for any irregular or complicated layouts, you will need to hand-tune the basic code an editor produces. You should therefore be familiar with all the tags related to tables, as well as their corresponding attributes and values. An HTML editor is most useful for quickly providing you with the "shell" of a simple table.

- *What is a good way to control the layout of a form?*

    Tables are the best way to exercise control over the exact layout of a form. For example, it is easy to create a columnar layout of radio buttons or checkboxes using a table. Other form elements can also be laid out precisely using tables.

- *Can I convert tables from other formatters, such as Microsoft Word, into HTML?*

    Yes, there are conversion utilities. However, in many cases, you will need to hand-tune the results.

EXERCISES 7.5 Tables

25. Create a tic-tac-toe game in the form of a table; show a win for X's.

26. Insert a three-by-three table into a long paragraph of text, position the table flush left, and position the text 10 pixels away from the table in both the vertical and horizontal directions.

27. Pick your favorite sport at school. Produce a table that lists the opposing team's name, their score, your team's score, and whether it was a W (Win) or L (Loss). The last row should "total" columns. Use results either from last year or the current season.

28. Generate a three-row, six-column table with the headers "**span 1-3**" and "**span 4-6**." These headers should span columns 1 through 3 and 4 through 6, respectively. The content of the table should be as follows:

- The first row has two items called "one-three" and "four-six." (As expected, they span columns 1 through 3 and columns 4 through 6.)

- The second row contains the single item "one-six."

- The third row has three items: "one-two," "three-four," and "five-six." (As expected, they span columns 1 and 2, 3 and 4, and 5 and 6, respectively.)

29. Produce a three-row, three-column table in which each column takes up 25 percent of a browser's window.

30. In a three-column table, what is the effect of setting the table WIDTH attribute to 75% and the table data WIDTH attributes all to 33%?

31. Generate an HTML table that describes your class schedule this semester. Include a caption beneath the table.

32. Does your browser support the BORDERCOLOR, BORDERCOLORDARK, or BORDERCOLORLIGHT attributes for any of the following tags: table, table data, table header, or table row? Experiment with these attributes and, based on your findings, describe their functions.

33. Program a table that is borderless and contains the colors of the rainbow as columns.

34. What is the function of the NOWRAP attribute of the table data and table header tags?

35. Find an issue of your campus newspaper. Lay out its front page on-line using tables. (Provide descriptions of the graphics.)

36. Welcome to the halls of industry. Your boss walks in and says to you, "I need an HTML table with the following specifications and I want it now!" She hands you a sheet of paper and walks out of the room saying, "I'll be back in 10 minutes." The paper says:

> Produce the HTML code for the following table: it is three columns wide and two rows deep. From left to right, top to bottom, the colors should be red, orange, yellow, green, blue, and salmon. The contents of the cells, in the same order, are to be Songs, Books, Recipes, Junk Mail, Autos, and Flowers. The contents are to be centered, and some space, say 20 pixels, should surround the text in each box. The columns should each take up exactly 25 percent of the browser's window. Don't include borders if you'd like to keep your job!

Reading the specification has taken longer than you anticipated. However, you feel inspired. Code the table described in the specification and display it on the Web.

7.6 Debugging

Using examples, we review the most common mistakes made by HTML programmers, as well as how to avoid them. For hard-to-find bugs, you can always
resort to an HTML syntax checker. As you will see, the key to minimizing the
time spent during the debugging phase is to use a good programming style.

Two of the most common mistakes in HTML programming are:

1. Omitting the leading / on an ending tag, thereby not closing its corresponding beginning tag, as intended, but instead opening a new environment.

2. Incorrectly nesting tags.

By using an appropriate programming style, you can avoid both of these problems, or at least make them easier to detect.

To illustrate the first type of error, we present two examples of flawed
HTML code that center, emphasize, and increase the font size of the word
"Warning."

```
    ...
<CENTER>
    <STRONG>
        <FONT SIZE = "+2">
            Warning
        </FONT>
    </STRONG>
<CENTER>                <!-- Bug -->
    ...
```

Because blank lines separate this code from its surrounding code (indicated
by the "..." in the example) and because the tags are indented, it is easy to see
that the second center tag is missing the leading / symbol. Contrast this code
with the following:

```
<TABLE BORDER=2><TR><TD>TR test</TD></TR>
</TABLE><CENTER><STRONG><FONT SIZE=+2>
Warning</FONT></STRONG><CENTER><EM>compressed
code is hard to read</EM>
```

It is much more difficult to detect the problem with the center tag in this case,
because the code is jammed together.

The next example illustrates nesting errors:

```
...
<CENTER>
    <STRONG>
        <FONT SIZE = "+2">
                Warning
        </FONT>
    </CENTER>                              <!-- Bug -->
</STRONG>
    ...
```

It is easy to see that the beginning strong and ending center tags have been mismatched. Contrast this with the following code:

```
<TABLE BORDER=2><TR><TD>TR test</TD></TR>
</TABLE><CENTER><STRONG><FONT SIZE=+2>
Warning</FONT></CENTER></STRONG><EM>compressed
code is hard to read</EM>
```

It is obviously much more difficult to detect the bug in this case. If you indent your code, use blank lines, and use consistent capitalization in tags, your debugging will be considerably easier.

EXERCISES 7.6 Debugging

For each problem in this set of exercises, indicate the bug(s), if any; otherwise report "no bug."

37. `<PARAGRAPH> Welcome to my Homepage. </PARAGRAPH>`

38. `<h6>Welcome to my Homepage.<h6>`

39.  ```
     <UL>
         <DD> Data goes here.
         <DD> More data goes here.
     </UL>
     ```

40. `<EMP> italics </EMP>`

41. `<STRONG>this would probably be boldface<STRONG/>`

42. `<HEADER> Red Hot Dollar </HEADER>`

43. `<H3 ALIGN == "almost-center">Just Off</H3>`
     ```

44. `<FONT WIDTH = "10%" COLOR = "red">Red text</FONT>`

45. `<IMG SOURCE = "bruins.gif" ALGN = "left">`

46. `<HEAD> <TITLE> Welcome Yoda! </HEAD> </TITLE>`

47. `<BODY BACKGROUND = "white">`
    `        Under construction.`
    `</BODY>`

48. `<HR COLOR = "blue">`

49. `<h4> Small text. </h4<`

50. `<SMALL <SMALL <SMALL tiny text goes here >>>`

51. `<OL BEGIN = "3">`
    `        <LI> Number 3.`
    `        <LI> Number 4.`
    `</OL>`

52. `<BODY BGCOLOR = "#FGAA00"></BODY>`

53. `<A HREF = "http://~petreschi\myfile.html"> Rossella</A>`

54. `<ALIGN CENTER> In the middle of the road. </CENTER>`

55. `<ITALICS> in slanted form </ITALICS>`

56. `<IMAGE = "glass.html">Soft drink</IMAGE>`

57. `<BR> Make a new line here. </BR>`

58. `<IMG SRC = "me.gif"`
    `        WIDTH = "10 pixels"`
    `        HEIGHT = "20 pixels">`

59. `<OL STYLE = "Roman">`
    `        <LI> Florence`
    `        <LI> Naples`
    `        <LI> Venice`
    `        <LI> Verona`
    `</OL>`

60. `<CAPTION = "Data for the Heart">`
    `<TABLE>`
    `        <TH> Time`
    `        <TH> Number of Beats`
    `</TABLE>`

61. `<HTML>`
    `<BODY>`

```
<BODY BGCOLOR = "orange">
<TITLE>
<B>My Orange Tree Page</B>
</TITLE>
You will learn all about citrus fruit
on this page.  In particular, my favorite ---
oranges.
</BODY>
</HTML>
```

62.
```
<P ALIGN = "center>
<TABLE>
        <TR ROWSPAN = "3" COLSPAN = "4"
        <TD> Big item
        <TD> Small item
        <TR>
        <TD> Next item
<TABLE>
</P>
```

63.
```
<TABLE BORDERCOLOR = "green">
        BGCOLOR = "red"
        ALIGN = "center">
        <TC>
                <TD> Column 1
                <TD> Column 2
                <TD> Column 3
<TABLE>
```

7.7 More Advanced Topics

We have described a wide range of basic HTML features up to this point
in the text. However, there are many other interesting topics, such as Web
graphics, frames, forms, CGI scripts, dynamic documents, HTML tools, and
Web programming, that we are unable to cover here. For these and other
advanced topics the reader is referred to the references contained at the back
of the book or to *In-line/On-line: Fundamentals of the Internet and the World
Wide Web*, published by McGraw-Hill.

Internet Service Providers

A

A.1 Introduction

An *Internet Service Provider* (*ISP*) is a business that sells computer access to the Internet. If a connection to the Internet is not available through an account at work or school (or if these connections do not permit you to roam freely, due to *firewalls* or *filters*), you may want to consider an ISP.

An option that we will mention only briefly is that of subscribing to a commercial on-line service provider, such as *America Online* (AOL). These commercial providers are designed to be user-friendly and are intended for people with little or no computer experience. They usually provide their own interfaces and proprietary services, such as *bulletin/message boards*, *chat rooms*, and *news hosts*. Content may be monitored,[1] which is something an Internet veteran may resent. However, those with children who are planning to use the service to access the Internet may appreciate this feature. In the past, the major drawback to using a commercial on-line service was the expense, but the current rates are fairly competitive with ISPs.

A persistent problem of the large commerical providers is that their service is oversubscribed. Sometimes, too many users are trying to get on-line at one time; this results in frustrating busy signals and e-mail delays.

ISPs typically offer a no-frills, cheaper alternative to commercial providers. People who are wise in the ways of the Internet find such a service appealing. Also, a good ISP will make newcomers to the Internet feel comfortable by providing on-line support and initial guidance.

Once you have decided that an ISP is the way to get on-line, the question becomes how to choose from the thousands that are available. A plethora of information on selecting an ISP is available on the Web, but if you do not currently have access to the Internet, this is not a viable solution. In the next sections, we offer several suggestions on how to shop for an ISP.

---

[1] A recent *Prodigy* case showed that monitoring can legally backfire with increased liability if something slips through.

A.2 Questions to Ask

When searching for an ISP, try to obtain answers to the following questions:

- Does the ISP provide a local-access telephone number for your area? This is probably the most important question, since you will otherwise wind up paying the telephone company long-distance charges for every minute on-line. You can verify that the access number is local by checking the phone book or asking the operator. Also find out if the ISP has more than one local access number, in case the line is busy during high-volume times (usually between 5:00 p.m. and 9:00 p.m.).

- What rate plans are available? Some ISPs offer a flat-rate fee for unlimited access. They may also offer an option for a lower monthly fee that includes "free" usage for a limited number of hours, after which you are charged for each additional hour. If both options are available, find out if you can switch between rate plans without a penalty. Also inquire about any other fees, such as a start-up fee, or surcharges for being on-line at certain times during the day.

- Does the ISP offer a free trial period? This is a good way to find out if the ISP is right for you. Test the connection during the same times that you anticipate using the service.

- What is the ISP's "user-to-modem" ratio? In general, a ratio higher than twelve to one (twelve users for one modem) means you may get a lot of busy signals when you dial in.

- What is the modem connection speed for each of the ISP's local access numbers? The connection speed should be at least 14.4 Kbps (hopefully, 28.8). What is the connection speed between the ISP and the Internet? A *T1* or *T3* connection is desirable. A T1 link supports a data transmission rate of 1.5 Mbps; a T3 link supports a rate of 44.7 Mbps. Thus, a T3 link is a much higher-volume connection.

- What types of accounts does the ISP offer? They may have *SLIP*, *PPP*, and/or shell accounts. SLIP (*Serial Line Internet Protocol*) and PPP (*Point-to-Point Protocol*) accounts permit TCP/IP traffic over telephone lines using a modem, and they allow the use of graphical browsers. Shell accounts provide only text browsing capabilities. A knowledge of UNIX, *VMS* (*Virtual Monitoring System*), or other appropriate operating system is also necessary to use a shell account. Some ISPs throw in a shell account gratis. A novice would be very happy with a personal PPP account, though a SLIP account is not very different. (PPP is gradually replacing SLIP.) If possible, you should obtain a PPP account.

- What software (such as Web browser, mail client, newsreader, and so on) does the ISP provide for Internet access? Is their documentation adequate? Do they s-mail you the software on disk, or do they provide a place from which you can download it? Does the ISP provide technical support for this process? If you do not care for their choice of software, will they allow you to obtain and use your own software? If possible, try the software the ISP supplies before making any subscription decisions.

- How much disk space is provided for e-mail? Does the ISP provide any disk space for customers to create and publish Web pages? If they do, how much storage is available and is there a fee involved? What if you need more space? These questions might not seem relevant now, but they will be once you start accumulating HTML files and images. You will probably need a minimum of 2 megabytes (2 megs) of space to work comfortably.

- Does the ISP have a toll-free technical support number? When are they available? Technical support 7 days a week, 24 hours a day (called 24–7) is wonderful, but smaller ISPs often do not have the staff to supply it. Technical support via e-mail works well if the technical support personnel respond in a timely manner and if the problem you are having does not involve the login process. Ask current users about their experiences with the ISP you are considering.

- Does the ISP offer an access number that you can use when you are traveling out of town and want to get on-line? If so, what is the charge per minute? Again, this may not seem important now, but it could be in the future.

- How many newsgroups does the ISP carry? How long do they keep the postings? Do they carry the newsgroups you like, and will you have an adequate chance to read them?

- If you are interested in registering your own domain name, does the ISP offer domain name services? In order for you to do this, the ISP must provide static IP addresses.

- If you plan to run a business via your Web access, compare the ISP's business rate schedule versus their nonbusiness rates.

A.3 Desirable Features

Some of the desirable features provided by good ISPs are outlined here. The weight you place on each item will depend on your individual needs. This short list was culled from the questions in the previous section. You may wish to add or delete features, depending on how you are planning to use your Internet connection.

- Local-access phone numbers for dialing in (a necessity).

- Local-access phone numbers that are not busy, even during the most congested times of the day (5:00 p.m. to 9:00 p.m.).

- Toll-free telephone number for customer support, plus the flexibility to e-mail questions to technical support staff.

- A 28.8 Kbps or faster modem connection between you and the ISP; a T1 or T3 connection between the ISP and the Internet.

- Five to ten megabytes of space for your own personal Web pages.

- A rate plan that suits your needs. For example, if you only plan to be on-line for a couple of hours each month, you may save money by not opting for a flat-rate plan.

- Newsgroup access that includes the newsgroups in which you are interested. Postings should be retained for at least two weeks. (Some ISPs keep postings for months.)

- Positive recommendations from other ISP customers.

A.4 Connection Setup

To connect to the Internet via an ISP, the following equipment is necessary: modem, telephone line, and computer with TCP/IP networking software installed. A connection is obtained by the modem dialing up the access number provided by the ISP. Assuming you have a SLIP or PPP account, you supply a login userid (which was assigned to you or which you had previously selected) and a password (again, selected by you), and you are then connected directly to the Internet. SLIP and PPP permit the TCP/IP data to be transferred over the telephone lines. Certain Internet software tools must also be installed on your computer, including a Web browser, mail program, newsreader, and so on. You actually run these clients on your computer. For shell accounts, these clients are run on the *remote host*. As noted earlier, the ISP may make these clients available to you, or may allow you to select and install your own. Getting all of these components installed and configured may not be trivial, but it can be made much easier by both good basic information from the ISP and help from someone with experience.

A.5 Typical Problems

Despite your asking all the right questions and doing your ISP homework, problems may arise. Some problems are common among ISPs, and they can motivate people to switch ISPs.

One of the most frustrating problems associated with ISPs (and commercial on-line services) is the busy signal when you are trying to connect. This may occur only at certain times of the day, but if you need to get on-line, this situation presents a problem. Also, the problem may develop over time, as the

ISP grows and cannot keep up with user demand. While evaluating the ISP, we recommend dialing the access numbers and trying to connect, to determine how often the line is busy (if at all) at various times of the day. Talking to others who use the ISP can also provide insight on how difficult it is to connect. In addition, remember that your own telephone line will be busy when you are on-line, so you might have to consider getting a second phone line.

Another problem sometimes encountered is slow download times, even though your modem connection is fast. If your modem is 28.8 Kbps and the ISP's modem connection is 14.4 Kbps, data is transferred at the slower rate. You will also feel this effect if the ISP's connection to the Internet is slow, or if they have a lot of customers placing high demands on the system. If you read about 56 Kbps modems and have dreams of higher speeds, consider that the quality of your local telephone lines may limit you to 28 Kbps or lower. Talk to others in your area and to your ISP to find out their experiences.

ISPs to which you cannot connect create an annoying problem, especially since this may not become apparent until after you have subscribed. Similar problems can arise with an *e-mail only service*. E-mail messages may be lost, or may not be delivered in a timely manner. If your ISP addresses your problems right away, these issues may not be a concern. The best way to ascertain the ISP's reliability is talk to their customers.

Many users have difficulty configuring TCP/IP software for their systems. Sometimes, the setup documentation provided is not clear or is hard to follow. The installation process is usually not intuitive and is system-specific. If you encounter any technical difficulties, the best way for you to proceed is to contact the ISP's technical support staff immediately. When describing your problem, try to provide them with as much detail as possible.

A.6 Internet Service Provider Selection

To select an ISP, you need to find out which ISPs are in your area (remembering that you do not want to make a toll call to connect). If you currently have access to the Internet, you can check the Web for lists (and critiques, too) of providers in a specific area. There are also newsgroups that address ISP selection. Other alternatives include looking up ISPs in the off-line Yellow Pages or talking to your friends.

Once you have compiled your list of ISPs, contact them by phone or e-mail to get answers to the questions posed here. Some ISPs have Web presentations and visiting them would provide you with additional information. In fact, the appearance of their Web presentation might provide some insight into how thorough and professional they are. Whether the ISP's staff answers your e-mail inquiries quickly might provide clues about how promptly they will respond to your questions once you are a customer.

Using the answers to the list of questions as a guide, eliminate the ISPs that are not suitable. For example, one of them may not carry a newsgroup you always read, or another may not provide an access number for you to use when you travel. Next, if possible, test the ISP's access lines by using your modem to connect to theirs. Connect at different times of the day, to establish when and if busy signals are going to be a problem. Also, try calling their toll-free technical support number, if they have one. Note how long you have to wait until you speak with someone. Was the person who answered knowledgeable and friendly?

Once you have narrowed down your list, try to talk to other users about the candidate ISPs. Their comments may be the most useful pieces of information. If the ISPs offer a free trial period (or a money-back guarantee for a period of time), try them out. Keep a record of the research you have done up to this point, since it may come in handy later.

Once you have selected your ISP, unexpected problems may arise after you have used the service for awhile. If a problem becomes acute, you may have to switch to your second choice.

Our last piece of advice is to pay for the service by the month, not by the cheaper yearly rate (if offered). Then, if you want to change to another ISP, you do not have to wait a year, or get penalized by leaving earlier.

Text Editing

B.1 Introduction

This appendix contains basic information about *text editing*. It is intended for those who have little or no editing experience. A text editor is a program that is used to create and modify ASCII files. To introduce you to basic text editing commands and editing principles, we will use the *Pico* (pronounced PEE-ko) editor, but any simple text editor will suffice. Details about other specific editing software are readily available on-line.

Many different varieties of text editors are available. In addition, numerous word processing programs come with built-in text editors. Some text editors have graphical user interfaces, and many commands are mouse driven. Other editors are more keyboard-based. The decision as to which style of editor to use will depend in part on what is available to you. Hopefully, you will be able to experiment with a number of different editors and then select the one you prefer. Here, we focus on keyboard-based editors, as opposed to graphical editors. Graphical editors usually provide extensive on-line help and pull-down menus, with icons to indicate how to perform the various commands.

B.2 Keyboard Driven Editors

Two popular and powerful text editors are `vi` and `emacs`, which are a little too complex to be used in an introduction to text editing. Another, simpler text editor is called Pico, which is supplied with the *Pine mail program* developed at the University of Washington. Pico stands for *Pine composer*, and during message composition, the Pine mailer defaults to the Pico editor. The basic principles and editing features of Pico apply to other keyboard-driven editors.

B.3 Pico

This short tutorial will provide a new user with the minimum editing skills needed for file composition. Editing commands can be split into several groups: cursor movement, cut and paste, save and insert files, and miscellaneous. Most editors, including Pico, let you insert text at the current cursor position simply by typing it in.

Figure B.1 shows a screen shot of the Pico interface. Many of the commands are executed by depressing the CONTROL key and another key simultaneously. On any system, the command for obtaining help is very important. Pico help allows you to obtain documentation about all of the editor's commands on-line. To get help from inside Pico you type (2 keys)

    CONTROL-G

The G does not need to be upper case. In the figure, the control key is shown as a caret (∧). In our examples, we prefer to use the word CONTROL. We will describe each of the other commands listed at the bottom of the Pico screen and will follow the Pico convention of using all capital letters.

FIGURE B.1 The Pico Text Editor Interface.

```
─ nxterm                                                             _ □ ▣
   UW PICO(tm) 2.9                       File: welcome.html
 on-line is so vast that someone interested in obtaining timely news,
 stock updates, or basic research information needs to become a
 competent computer user.  There are discussion groups on the Web
 covering every topic imaginable.
 </P>

 <P>
 The level of sophistication we are aiming for in this book is not
 at the ``point and click'' level nor at the ``hacker'' end of the
 spectrum.  We are interested in helping computer users learn enough so
 that they are comfortable performing the following functions (among
 others):
 </P>

 <UL>
 []
 <LI>
 sending and receiving electronic mail (e-mail)

 <LI>
 browsing the World Wide Web

 <LI>
 publishing on the World Wide Web

 <LI>
 reading and posting to <EM>newsgroups</EM> and <EM>mailing lists</EM>

 <LI>
 submitting forms on-line

   ^G Get Help  ^O WriteOut  ^R Read File  ^Y Prev Pg  ^K Cut Text   ^C Cur Pos
   ^X Exit      ^J Justify   ^W Where is   ^V Next Pg  ^U UnCut Text ^T To Spell
```

B.3.1 Cursor Movement

The basic cursor movement commands are:

- CONTROL-F—Forward one space.

- CONTROL-B—Backward one space.

- CONTROL-N—Next line.

- CONTROL-P—Previous line.

- CONTROL-A—Beginning of line.

- CONTROL-E—End of line.

More specialized movement commands, such as CONTROL-Y to return to the previous screen and CONTROL-V to move forward a page, are also available. The command CONTROL-C reports the current cursor position on the screen. The arrow keys can also be used to reposition the cursor.

B.3.2 Cut and Paste

One of the most important editing operations is the ability to *cut* and then *paste* a piece of text. If done properly, this move can save considerable time. In Pico, the cutting is done using the CONTROL-K command, and the pasting is done using the CONTROL-U command. By repeatedly using CONTROL-K, you can "select" a number of lines to be cut. For example, five CONTROL-K's will cut five lines. Moving the cursor and then using a single CONTROL-U will paste the cut text into the place where the cursor is positioned.

B.3.3 Save and Insert Files

How do you insert a file into the file you are currently editing? This is done using CONTROL-R. When you execute this command, you will be prompted for the name of the file to insert.

How do you save a file? To save a file, use the CONTROL-O command. To exit the editor without saving a file, use the CONTROL-X command.

How do you create a new file? Enter the command Pico at the operating system prompt, and then edit and save the new file using the CONTROL-O command. Another way of doing this is to type

```
%pico new.file
```

where new.file is the name of the file you want to create. When you save the file, it will be written to new.file.

B.3.4 **Miscellaneous**

There are four important miscellaneous features. They are spell-checking, searching within a file, justifying text, and suspending the editor.

Spell-Check

Pico has a built-in spell-checker, which is invoked by the command CONTROL-T. If the spell-checker finds a word that it thinks is misspelled, it offers you the opportunity to correct the word. On most systems, other spell-checkers are available. It is a good idea to run your files through the spell-checking phase, but do not assume that all spelling errors have been corrected. The spell-checker will not catch words that are spelled correctly but are the wrong words. For example, if "there books" should be "their books," the spell-checker will not catch it.

Search

To search for text (or a string of text) within a file, use the CONTROL-W command. This command is very useful when you are entering edits from a hardcopy. The command allows you to search the file to get close to where you need to enter a change. Simple cursor movements can then bring you to the exact spot to edit.

Justify Text

The CONTROL-J command is used to justify the text of the current paragraph. (It does not allow you to set the margins within which justification takes place.) When composing e-mail, this is very useful for pretty printing a paragraph. That is, CONTROL-J makes the lines appear near equal in length.

Suspend Pico

If you invoke Pico with the -z flag,

```
%pico -z new.file
```

you will be able to suspend editing by typing CONTROL-Z; that is, using CONTROL-Z exits Pico to the shell prompt. This lets you look up or obtain information with other commands and then resume your editing session by issuing the command:

```
%fg
```

B.4 **Commands Summary**

A summary of Pico commands is as follows:

- CONTROL-A—Move cursor to beginning of line.
- CONTROL-B—Move cursor backward one space.

- `CONTROL-C`—Report current cursor position on the screen.
- `CONTROL-E`—Move cursor to end of line.
- `CONTROL-F`—Move cursor forward one space.
- `CONTROL-G`—Display Pico help.
- `CONTROL-J`—Justify text in current paragraph.
- `CONTROL-K`—Cut text.
- `CONTROL-N`—Move cursor to next line.
- `CONTROL-O`—Save file.
- `CONTROL-P`—Move cursor to previous line.
- `CONTROL-R`—Insert file.
- `CONTROL-T`—Invoke spell-checker.
- `CONTROL-U`—Paste text.
- `CONTROL-V`—Move forward a screen.
- `CONTROL-W`—Search for a pattern of text.
- `CONTROL-X`—Exit Pico without saving the file.
- `CONTROL-Y`—Move backward a screen.
- `CONTROL-Z`—Suspend Pico.

Pine Mail Program

C

C.1 Introduction

This appendix explains some of the features of the *Pine mail program*, a popular e-mail program developed at the University of Washington in 1989. The name "Pine" stands for *Program for Internet News and Email*. Pine provides an easy-to-use, keyboard-driven interface for composing, sending, reading, and managing e-mail and newsgroup messages.

We will explain some of the basic features of Pine. Other mailers have similar functionality, but their user interfaces may differ. For example, *Eudora* is a popular mouse-driven e-mail program. Nearly all features found in Pine exist in Eudora and vice versa. However, in Pine, one or two keystrokes are needed to execute a command, whereas in Eudora, several mouse clicks are necessary. (Both Pine and Eudora have extensive on-line documentation.) As with any software, the best way to learn about it is to get on-line and experiment.

C.2 Getting Started

Pine is text-driven; that is, all commands are entered as key strokes. The mouse is not used, as there is no graphical interface. Menus, located at the bottom of the screen, guide the user by displaying the various commands and options. On-line help is easily accessible, making Pine particularly user-friendly.

Pine software must be installed on your computer (or on the system of which your computer is part). Typing pine at the UNIX prompt, or selecting Pine from a menu of options, are two ways to start the program.

When Pine starts, the menu shown in Figure C.1 is displayed. This is the *Pine Main Menu*, which lists Pine's features/options. The top line of the screen indicates the version of Pine that is running (in this case, 3.96). The title of the screen, MAIN MENU, appears next to the version, and to the right is a message indicating how many messages are there. In Figure C.1, we see that there are six messages in the folder named INBOX.

FIGURE C.1 Pine Main Menu Screen.

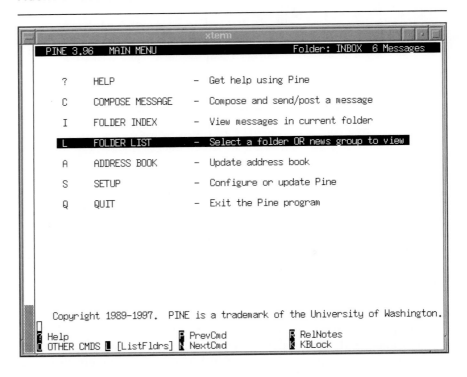

The features available, as listed on the screen, along with their associated commands, are as follows:

• Getting on-line help about Pine—?.

• Composing a message—C.

• Viewing messages—I.

• Selecting a folder of messages to view—L.

• Updating the address book—A.

• Customizing Pine—S.

• Exiting Pine—Q.

Notice in Figure C.1 that in addition to listing the options with their descriptions, the screen also offers a summary of commands along the bottom. Since all of the available options cannot fit at the bottom, pressing the O key will display "other commands" that are available. Some of the commands displayed include P (or PrevCmd) for "previous command" and N (or NextCmd) for "next command." Using the P (or N) key highlights the previous (or next) option in the features list. A highlighted option can then be selected by pressing the return key.

Another way to select an option is to press the associated key, either uppercase or lowercase, without pressing the return key. Case is not significant. Pine itself displays commands in uppercase letters, and we have adopted that convention. For example, pressing the C[1] or c key allows you to compose a message, and pressing the Q key allows you to quit Pine. (When using a conventional mouse, only one hand remains on the keyboard. However, one advantage of Pine is that both hands may remain on the keyboard, so you are always in a position to type. Not having to reposition the mouse cursor and the other hand on the keyboard can let a skilled typist operate more efficiently.)

The R at the bottom of the Main Menu screen stands for "Release Notes." The Release Notes describe any changes from the previous version of Pine, bugs, and so on. The K command permits you to lock your keyboard, for those times when you want to step away from your computer for a few minutes, but do not want to log out and do not want anyone else to use your computer.

In addition to the summary of commands and the list of options, the Main Menu screen provides an area for messages and prompts. They will be displayed in brackets just below the copyright/trademark declaration. For instance, entering D on the Main Menu screen results in the following message being displayed in the message/prompt line:

```
[Command "D" not defined for this screen. Use ? for help]
```

C.3 Composing and Sending Mail

To compose an e-mail message, press the C key (or select the COMPOSE MESSAGE option by highlighting it and pressing the return key) from the Main Menu. A new screen titled COMPOSE MESSAGE will appear, containing a *template* for constructing the message. Figure C.2 depicts that template.

Like the Main Menu, the bottom of the Compose Message screen contains a menu of pertinent commands. These commands show a ∧ character before a letter, such as ∧G. The commands are executed by pressing the control key (Ctrl) at the same time as the letter key, either uppercase or lowercase. In our discussion, we use the word CONTROL to refer to these commands; that is, we use CONTROL-G, although what you will see on the screen and in our figures will be ∧G.

The CONTROL-G (Get Help) command found on the bottom of the screen provides help information about whichever template field is highlighted. The

---

[1] Pressing the C key is shorthand for typing SHIFT-c.

FIGURE C.2 Pine's Template for Message Composition.

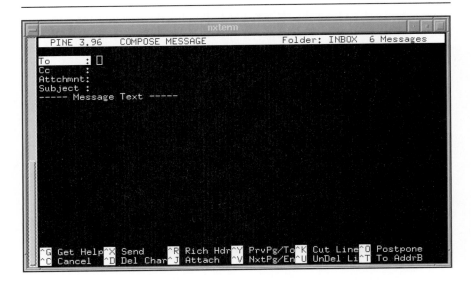

basic template includes a To: field, Cc: field, Attchmnt: field, Subject: field, and Message Text area. Use the arrow keys (←, →, ↑, or ↓) or CONTROL-P (previous line) and CONTROL-N (next line) to move from field to field.

You will notice that as different fields in the template are highlighted, the commands at the bottom may change. For instance, when the To: field is highlighted, selecting the CONTROL-T option provides access to a list of addresses in the *address book*. When the Message Text field is highlighted, CONTROL-T invokes the spell-checker. Pine provides the help menus at the bottom of each screen.

Text is entered in the various fields using the Pico text editor. More information about Pico can be found either in Appendix B or on-line while using Pine. When you highlight the To: field and enter the CONTROL-G command, the help screen provides a short summary of Pico cursor movement and editing commands.

C.3.1 The To: Field

When composing an e-mail message, you enter the address of the recipient in the To: field. Position the cursor at the To: field, which is then highlighted, and type in the e-mail address. More than one recipient can be specified by separating them with commas. (This is called *comma delimiting*.)

A nickname or alias can be specified in the To: field. Pine will check your address book to determine the full e-mail address(es) and will expand the nickname for you. (Recall that an alias can specify a list of e-mail addresses, as well as a single address.) Nicknames can be specified, along with regular e-mail addresses, when more than one recipient is listed. For example,

```
To: dickens@aol.com, marty, club
```

The nickname "marty" is an alias for mphillips@biology.uchio.edu, and "club" is an alias for a list of e-mail addresses belonging to the members of the "Breakfast Club."

C.3.2 The Carbon Copy Field

The Cc: field allows you to send copies of a message to other people. E-mail addresses, as well as nicknames, can be entered in the field. This field is optional and can be left blank either by skipping over it (CONTROL-N) or by pressing the return key when the cursor is positioned at it.

C.3.3 The Attachment Field

The Attchmnt: field lets you specify the name of files to append (or "attach") to a message. Attachments are appended separately and are not part of the message text. Only files residing on the same machine as the mailer can be appended to the message, and only recipients with MIME-capable mail programs (such as Pine) will be immediately able to view attachments that are other than just plaintext. When the cursor is positioned at the Attchmnt: field, the CONTROL-T command displays files from which to select. After entering the name of the file, or using CONTROL-T to select a file, you simply press the return key to attach the file.

C.3.4 The Subject Field

The Subject: field provides a place for you to tell the recipient what your message is about. Enter a short, clear description of your message.

C.3.5 Other Header Fields

Entering the CONTROL-R (Rich Hdr) command while in the Compose Message screen causes your template to display other optional fields. To use one of these fields, simply fill in the appropriate information. The Bcc: field permits you to send "blind carbon copies," that is, the addresses of the copied recipients are not displayed. The Fcc: field lets you specify a folder in which to keep a copy of the outgoing message. Other fields, including Newsgrps: and Lcc:, can be explored by using the CONTROL-G command.

FIGURE C.3 A Sample Pine Compose Message Screen.

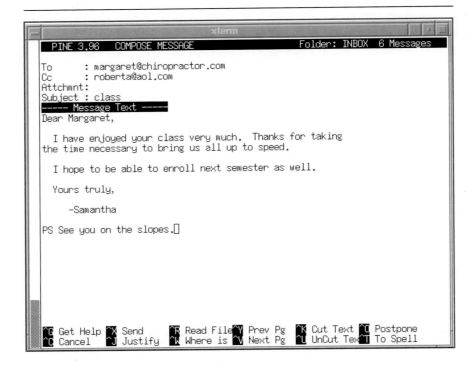

C.3.6 Message Text Field

To enter the body of your message, position the cursor over the Message Text field. Using Pico text editing and cursor movement commands, type your message. When you are done, invoke the spell-checker by pressing the CONTROL-T. The CONTROL-R command lets you insert a file into the message.

Once your message is complete, you can either send the message using the CONTROL-X command, cancel the message with the CONTROL-C command, or postpone sending the message until a later time, using the CONTROL-O command. Each of these commands causes Pine to ask you if that is what you really want to do. You respond by pressing either the Y key for "yes" or the N key for "no." Pressing the return key will invoke the default selection, which is shown with brackets around it. Figure C.3 illustrates a Pine Compose Message screen that is ready to be sent.

C.4 Reading Mail

When you receive e-mail messages, Pine places them in a folder called INBOX. To view the messages in your INBOX folder, press I (Folder Index) from the Main Menu, which brings up a screen called Folder Index. This screen displays information about each message in your INBOX folder. Figure C.4 shows a sample INBOX folder screen.

The Folder Index screen does not display the full text of the messages you have received; instead, it displays specific information about each message, so that you can decide which (if any) messages you want to read. In Figure C.4, four messages are shown on the screen. This is also indicated on the top line of the screen, where it identifies the folder as INBOX and the highlighted message as 4 of 4.

For each message displayed on the Folder Index screen, seven pieces of information are provided:

Column 1—Either a blank or a "+" symbol, where a + indicates that this message was sent directly to you (as opposed to being carbon copied, for instance).

Column 2—Message status; either A, D, or N. An A means that the message was answered using the Reply command, a D signifies that you have read the message and have marked it for deletion, and an N indicates that the message is new.

FIGURE C.4 A Pine Folder Index Screen.

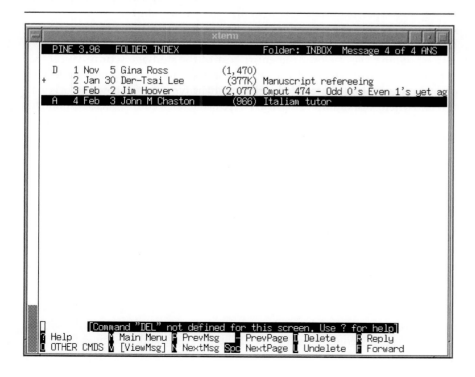

Column 3—Message number.

Column 4—Date the message was sent.

Column 5—Sender of the message.

Column 6—Size of the message, in bytes.

Column 7—Subject of the message, as specified by the sender.

To read a particular message, select it by using the arrow keys or by pressing the P (PrevMsg for Previous Message) key or the N (NextMsg for Next Message) key. The selected message is highlighted in reverse video and indicated on the top line: Message 4 of 4, for instance. Pressing the V key displays the entire message on another screen (perhaps replacing the original screen) called the Message Text screen. Simply pressing the return key does the same thing, since V is the default. The commands to continue reading messages are listed on the bottom of the screen: N (NextMsg) to view the next message, or P (PrevMsg) to view the previous message in this folder. To return to the index of messages, press I.

After reading a particular message, you might decide to respond to it. One way to do this is to go to the Compose Message screen, fill in the header information, write the message, and send your response. Another way is to use the Reply command, R, from the Message Text screen. From the message/prompt line, Pine will ask if you want to include the original message in the reply. Responding "no" (the default) causes the Compose Message Reply screen to appear with the header information already filled in. The Subject: field will contain Re:, followed by the subject line of the original message.

Responding "yes" causes the entire text of the message sent to you to be copied into the message text area, with header information filled in as well. The copied message text is preceded by a line identifying who wrote the original text and when. For example, if Enrico were using the R command to respond to a message sent by Laura (verona@books.com), Pine would put the following into the Message Text area:

On Sun, 18 Jan 1998, verona wrote:

Laura's original message would appear after that, with each line from her original message marked by a > symbol along the left margin. Lines can be added, deleted, and modified using the Pico text editing commands.

After completing your reply, you would use the same commands that appear on the Compose Message screen to finish: CONTROL-X to send the message, CONTROL-C to cancel the message, or CONTROL-O to postpone sending the message.

The Forward command (F) is similar to the Reply command. It allows you to forward the message you are currently reading to another e-mail address. A screen appears containing a copy of the message, and you must fill in the

To: field. Like the Reply command, the original message can be modified and added to.

C.5   Managing Mail

Managing e-mail involves deleting unwanted messages and organizing those messages you want to save. The saved messages can include copies of e-mail that you have sent, as well as mail you have received. Messages are organized and kept in *folders*. Pine provides the following three default folders for each account:

- INBOX   The folder where new incoming messages are stored.

- saved-messages   The default folder for saving copies of messages.

- sent-mail   The folder where copies of messages that you have sent are saved.

Pine allows you to create your own folders, in addition to these three.

To move around between folders, press L (FOLDER LIST) from the Main Menu. A Folder List screen appears, showing the *collections of folders*. To select a collection, highlight the phrase

"[Select Here to See Expanded List]"

below the desired collection and press the return key. The folders within that collection will be displayed on a new Folder List screen (possibly replacing the original screen), and you can highlight a specific folder to view.

A menu of commands is shown at the bottom of the screen. These let you: view what is in the selected folder (V—ViewFldr), delete (D—Delete) folders, add (A—Add) folders, and rename (R—Rename) folders. Another way to access the index of another folder is to use the G (GotoFldr) command, which prompts you to enter the name of the folder you want to view.

C.5.1   Deleting Messages

It is important to delete unwanted messages, since they clutter up folders and make it difficult to keep track of more important correspondence. Also, most e-mail accounts only have a limited amount of memory allocated to them, so cleaning out folders is necessary to conserve disk space.

Deleting a message requires two steps. First, the message is marked for deletion by selecting the message from the Folder Index screen. Using the D command marks the message for deletion. A message can also be marked for deletion from the Message Text screen, when you are viewing the message.

If you decide later that you do not want to delete the message, unmark the message using the undelete command, U, from either the Folder Index screen or the Message Text screen.

Second, *expunge* the messages that have been marked for deletion. You can deliberately expunge a message from the Folder Index screen by using the X command. Otherwise, when you terminate Pine or when you leave a folder (other than the INBOX folder), Pine asks if you want to expunge messages marked for deletion. Note that a message marked for deletion can be undeleted. However, once a message is expunged, it is gone for good and Pine cannot restore it.

C.5.2 Saving Messages

Incoming messages are automatically put in your INBOX folder. After reading a message (and possibly responding to it), you may decide to save it. If you are in the Message Text screen, pressing S saves a copy of the message and marks the INBOX copy for deletion. The saved-messages folder is the default folder in which messages are saved, but you can specify another folder. After you press S, Pine prompts you with:

```
SAVE to folder in <mail/[]> [saved-messages]:
```

Pressing the return key selects the default folder (saved-messages), or you can enter another folder name. If the folder name you specify does not exist, Pine asks if you want it created. Pine then saves the message to the file specified.

Messages can also be saved from the Folder Index screen by selecting the message (i.e., highlighting it) and then using the S command, in the same way as messages are saved from the Message Text screen.

C.5.3 Additional Features

While we have covered the most important features of Pine, the software includes many other features. To explore these features, we encourage you to experiment with Pine, especially the S (SETUP) command from the Main Menu, which will, among other things, allow you to create a signature file. Exploring the various commands and trying them out—sending mail to yourself, if necessary—is the best way to become proficient at using Pine.

Basic UNIX

D

D.1 Introduction

UNIX is a widely used operating system, particularly in academic and research settings. In addition, many operating systems have adopted UNIX-like commands and features.

In Section 2.5, we explained how to install a Web page on a UNIX-based Web server. (Installation of Web pages on Windows-based servers or other types of servers requires a different procedure.) In this appendix, we describe basic UNIX commands and the UNIX file structure. Our aim is to provide an overview, so that you will feel comfortable creating and manipulating files for your Web page.

D.2 Sundry Facts

When using a UNIX account, you will be greeted by a *prompt* character. The default prompt is usually the % symbol, indicating that the system is ready for you to enter a *command*. The default is also called the *shell* prompt, because you are interacting with the outer layer (interface, or shell) of the operating system. The prompt symbol % may be changed, and many users change it to something like

```
computername>
```

where `computername` is the name of their machine.

Commands are instructions you give the computer. While there are hundreds of UNIX commands, about a dozen will suffice for our purposes. Fortunately, UNIX also provides an on-line manual that can be accessed using the `man` command. As long as you can recall a command name, the on-line manual can tell you how that command is used, as well as the *options* and *arguments* associated with it. For example, typing

```
%man man
```

provides you with on-line help about the `man` command.

Options to commands are typically specified by single-character *flags* prefaced by a dash (-) symbol. For example,

```
%man -k time
```

requests manual information about commands that have something to do with the keyword `time`. Notice the single blank spaces between the individual items in this command line. The `k` is a flag to the `man` command. The word `time`, as used here, is an argument to the command. Arguments are values that are passed to the command.

Normally, you will type in a command at the prompt, along with associated options and arguments. As an example, the command to list a *directory*'s contents is `ls`. (Think of a directory as a *folder*, if the folder terminology is more familiar to you.) When you type the command

```
%ls
```

the resulting output might look as follows:

```
calendar      classes      finances      letters
misc
```

By adding flags to the `ls` command, you can request additional information about the directory's contents, or format the output in a particular way. To list a directory's contents in the "long form," which provides associated properties of the files and subdirectories contained within it, use the `ls -l` command.

So far, we have seen that `man`, stands for manual, and `ls` represents list. Many UNIX commands are *mnemonic*. Since most commands are only two characters, some of the mnemonics are not terribly helpful for new users.

Hopefully, the basic style of UNIX commands is now clear. Their general form is

```
%"mnemonic command name" -flag(s) argument(s)
```

Multiple flags are written adjacent to each other with no intervening white space, whereas multiple arguments are separated by blank spaces.

We will now look at the UNIX file structure.

D.3 File Structure

It is important to understand how files and directories are organized under UNIX, so that you will be able to access them. The purpose of this section is to describe this structure.

Among other things, UNIX commands are used to create, manipulate, and change permissions on files. Many commands take a file name as an argument. The exceptions include the `date` command, which is used to report the current date and time.

Directories and files are organized in a tree-like hierarchy. The directory at the top of the hierarchy is known as the *root directory* and is represented by a forward-slash character, `/`. On a large UNIX system, some standard subdirectories under the root directory are `etc`, `usr`, and `tmp`.

Under normal circumstances, if you have an account on a UNIX system, you will begin your login sessions in your assigned *home directory*. You can use the tilde symbol (~) to refer to your home directory. Starting from the root directory and proceeding through the hierarchy until you reach your home directory results in a *path* to your home directory.

When the directory names you traverse through are concatenated together, the result is a *pathname*. Like URLs, pathnames can be absolute, called *full* in UNIX jargon, or relative, which is the same term used in HTML. The first forward slash in a pathname represents the root directory. Additional forward slashes in a pathname separate the names of the subdirectories. As an example, Figure D.1 shows Mary McCarthy's directory and file structure.

The UNIX file system has some special files (and directories) that are called *hidden files* (respectively, *hidden directories*) or *dot files*. The names of these files begin with a period symbol (.), such as `.cshrc`, `.login`, `.newsrc`, and `.pinesrc`. The `rc` stands for "run command" and is a historical artifact of some old operating systems. These hidden files usually contain configuration or initialization information, and it is not advisable to edit these files unless

FIGURE D.1 Mary McCarthy's Directory and File Structure.

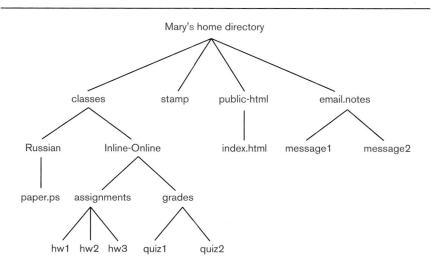

you are absolutely sure of what you are doing. Even then, you should make a backup copy of any such files you intend to edit. Some of these files are not intended to be edited by hand, but only by other programs. To list the hidden files and directories, you can use the -a flag (which stands for "all") to the ls command. Since you would rarely want to see or edit these files, the default for the ls command is not to list them.

D.4 File and Directory Organization

You have total control in creating and maintaining both files and directories in your home directory. One reason for using subdirectories is to organize files. In Figure D.1, for example, we see that Mary has subdirectories for classes, e-mail correspondence, and Web material. She also has a file called stamp in her home directory.

To create a subdirectory in any directory, use the mkdir command and provide the subdirectory name as an argument to this command. For example, to create a subdirectory called job.apps in the *current directory*, you would enter

```
%mkdir job.apps
```

The phrase "current directory," also called the *working directory*, refers to the directory in which you are working.

To move to another directory, you use the *change directory* command, cd. The command by itself, with no arguments, will place you in your home directory. To verify a change of directory, you can use the *print working directory* command, pwd. The output of the pwd command is the full pathname of the directory in which you are working. Suppose Mary just logged into her account named marymc and executed the command

```
%pwd
```

For output, she might see something like /export/home/marymc.

By specifying a pathname as the argument to the cd command, you can change directories. For example, to go into her Web directory, Mary can enter

```
%cd /export/home/marymc/public-html
```

Notice that a full pathname is specified in the command. Typing in a full pathname every time you want to change directories is time consuming; it is often more efficient to use a *relative pathname*. Using relative pathnames when navigating through the file structure can save a lot of typing time, not to mention potential typing errors. To move up one level in a directory hierarchy, follow the cd command with two dots

```
%cd ..
```

The two dots represent the *parent directory* for the directory in which you are currently located. To move from within her `public-html` directory to her `grades` directory, Mary can enter

```
%cd ../classes/Inline.Online/grades
```

Note that UNIX is case sensitive. In particular, some commands have both lowercase and uppercase flags that involve the same letter yet have totally different meanings.

D.5 File Manipulation

Once you know how to move around in the file structure, you are ready to create and edit files. Appendix B provides a short introduction to text editing for file creation. Here we concentrate on file manipulation. At some point, you will want to move a file to another directory, delete a file or directory, make a copy of a file, or just view a file. Numerous commands are used for manipulating files. One is the mv command, which is used to rename or move a file. To rename a file, follow the mv command by two arguments; the first is the name of the file you wish to rename, and the second is the new name for that file. For example,

```
%mv index.html index.old
```

renames the file `index.html` to `index.old`. After the command is executed, the file named `index.html` will no longer exist.

Another useful command is rm, which is used to delete or remove a file. The rm command should be used with care, because once the file is deleted, it is really gone—there is no undo. Using the -i option with the rm command will cause the system to ask for verification before deleting the file. We encourage you to use this flag. For example, entering the command

```
%rm -i mystuff.txt
```

will generate a query about whether or not you really want to delete the file `mystuff.txt`. The `rmdir` for deleting an empty directory is analogous to the rm command for deleting a file.

The `more` command can be used to look at the contents of a file. For example,

```
%more index.html
```

displays one screenful of the file `index.html`. Hitting the space bar brings up the next screenful of text, and typing q "quits" the command and brings you back to the prompt. The `more` command only allows you to view the file; you need a text editor to alter the file's contents.

The `cp` command is used to copy a file. This is different from the `mv` command in that you end up with another exact copy of the file; the original file and its name remain intact. As an example,

```
%cp index.html mypage.html
```

copies the file `index.html` to the file `mypage.html`. By using copies of HTML files that have similar tags, you can save considerable keyboarding and editing time. These copies become *templates* of the common HTML file formats on your Web presentation.

D.6   File Permissions

Permissions provide a measure of security by establishing who is able to access what files and directories, and how they can access them.

Three identifiers can be used to specify who can access an item. They are:

- The user who owns a file, designated u.

- Users who are members of the same group as the file owner, designated g.

- All other users, designated o.

Coincidentally, UNIX provides three different levels of file permissions, as well. The permissions, which are as follows, indicate how items can be accessed.

- Read, designated r.

- Write, designated w.

- Execute, designated x.

Their meanings are self-explanatory. To permit the use of its contents on Web pages, a subdirectory must allow both x and r access. With UNIX, write permission also means deletion ability. That is, if someone has write permission to a file (or directory), they also have the capability of deleting the file (or directory).

File and directory permissions can be conveniently displayed using the `ls -l` command. This command results in a listing of all files and subdirectories, along with their permissions, as given by a string of ten characters. Suppose the current working directory has just one file, called README. The command

```
%ls -l
```

might generate a response such as

```
-r--r--r-- 1 marymc Spanish 167 Jul 27 08:15 README
```

The first character identifies the type of item. In this case, a dash is displayed, which means an ordinary file. The other important type is directory, which is represented by a d.

Let us explain the other nine characters in the coding. The following graphic is helpful in visualizing these characters:

$$\underbrace{rwx}_{u}\ \underbrace{rwx}_{g}\ \underbrace{rwx}_{o}$$

The graphic shows that the remaining nine characters are split into three sets of three characters each. The first three characters are associated with user, the second set with group, and the last set with other.

For each set of three characters, an r in the first position indicates that the item can be read by members of the corresponding category. Similarly, a w in the second position indicates write permission, and an x in the third slot indicates execute permission. If any of the permissions are denied, the corresponding symbol is replaced by a dash. In this way, it is possible to have any combination of read, write, and execute permissions for each category.

For example,

```
-r--r--r-- 1 marymc Spanish 167 Jul 27 08:15 README
```

means the user, group members, and others can read the file named README; that is, everyone can read this file. In contrast, the permissions

```
-r-------- 1 marymc Spanish 167 Jul 27 08:15 README
```

indicate that only the user can read the file. Finally, the permissions

```
-rw-r----- 1 marymc Spanish 167 Jul 27 08:15 README
```

mean the user can read and write the file, group members can read the file, and everyone else is denied access.

In each of the examples presented here, the 1 in the second column indicates the number of *hard links* to the file. The user of the file is identified in the third column as `marymc`. The group is identified in the fourth column as `Spanish`. Finally, the 167 in the fifth column tells us that the size of the file README is 167 bytes; that is, it contains 167 characters.

As explained in Chapter 2, certain file and directory permissions are required so that others can read your Web pages. To set the permissions, UNIX provides the `chmod` command, which stands for change mode. There are two ways to use the command; numerically and symbolically. We will only describe the symbolic method here.

The `chmod` command takes as arguments a permissions setting and the category to be altered. Let us look at a couple of examples to clarify the use of this command. Suppose you execute an `ls -l` command and receive the following listing:

```
-rw-r----- 1 marymc Spanish 167 Jul 27 08:15 README
```

To add read permission for others, you would use the following:

```
%chmod o+r README
```

In other words, to add read permission for others, we specify others using `o` and use the plus (+) sign followed by an r flag. Similarly, to remove permission, we use the minus[1] (-) sign. For example, to remove the write permission on README, we enter the following command:

```
%chmod u-w README
```

An `ls -l` now results in the following:

```
-r--r--r-- 1 marymc Spanish 167 Jul 27 08:15 README
```

For illustration purposes, suppose we need to add write permission for the user and group, and execute permission for the user, group, and other. We can do this in two steps:

```
%chmod uog+wx README
```

followed by

```
%chmod o-w README
```

Notice that the first command temporarily gives write permission to everyone.

---

[1] This is the same symbol as dash, but we call it minus to indicate the removal of permission.

To change permissions on all items in a directory to world readable, we could do the following:

```
%chmod uog+r *
```

The star (*) is a wild card that equates to every item in the directory.

It is always good practice to check the file protections after using the chmod command, just to make sure they are really set the way you intend.

D.7 UNIX Command Summary

We have now provided sufficient background for you to install your Web pages on a UNIX server. For easy reference, we list the commands presented in this appendix, along with their functions:

- cd   Change directory.
- chmod   Change file permissions.
- cp   Copy a file.
- ls   List the contents of a directory.
- man   Display an on-line manual page.
- mkdir   Make a directory.
- more   Display a file one screenful at a time.
- mv   Rename a file.
- pwd   Print the working directory.
- rm   Delete a file.
- rmdir   Delete an empty directory.

HTML Tags

This appendix contains a convenient reference list of the HTML tags discussed in this book, along with a brief description of the function of each tag. The tags are listed in alphabetical order. The ellipsis symbol ($\cdots$) represents the material to be included between the beginning and ending tags. This appendix can be used in conjunction with the index to locate sample uses of the tags and additional information.

- `<A>` $\cdots$ `</A>`
  - Anchor tag, used for setting a hyperlink.
  - Attributes: `HREF`, `NAME`, and `TARGET`.
- `<ADDRESS>` $\cdots$ `</ADDRESS>`
  - Semantic based style type.
  - Used to indicate an address.
- `<B>` $\cdots$ `</B>`
  - Syntactic based style type.
  - Used to place text in boldface.
- `<BASE>` $\cdots$ `</BASE>`
  - Used for setting global parameters.
  - Attributes: `HREF` and `TARGET`.
- `<BASEFONT>` $\cdots$ `</BASEFONT>`
  - Used for specifying a document's font size.
  - Attribute: `SIZE`.
- `<BIG>` $\cdots$ `</BIG>`
  - Semantic based style type.
  - Used to increase the relative font size of the text.
  - Produces the opposite effect of the small tag.

- `<BLINK>` ⋯ `</BLINK>`
  - Syntactic based style type.
  - Used to create blinking text.
  - Use sparingly, since some users find the effect annoying.
- `<BODY>` ⋯ `</BODY>`
  - Indicates the start of the main part of an HTML document.
  - Attributes: `ALINK`, `BACKGROUND`, `BGCOLOR`, `LINK`, `TEXT`, and `VLINK`.
- `<BR>`
  - No associated ending tag.
  - Used to force a line break.
  - Attribute: `CLEAR`.
- `<CAPTION>` ⋯ `</CAPTION>`
  - Used to annotate a table.
  - Attribute: `ALIGN`.
- `<CENTER>` ⋯ `</CENTER>`
  - Syntactic based style type.
  - Centers whatever is enclosed between the beginning and ending tags.
- `<CITE>` ⋯ `</CITE>`
  - Semantic based style type.
  - Used to specify a reference.
- `<CODE>` ⋯ `</CODE>`
  - Semantic based style type.
  - Used to specify program code in the text.
- `<!-- ⋯ -->`
  - Comment tag, used for including notes to yourself in an HTML document.
  - Useful for including copyright notices in files.
  - Comments are not rendered by the browser.
- `<DD>` ⋯ `</DD>`
  - Used to identify the definition portion of an item in a "Definition List."
  - `DD` stands for "Data Definition."
  - Ending tag is usually omitted.

- `<DIV> ⋯ </DIV>`
  - Used to surround a group of HTML tags to control their alignment.
  - Attribute: `ALIGN`.

- `<DL> ⋯ </DL>`
  - Used to specify a "Definition List," where each item in the list consists of a term and its corresponding definition.

- `<DT> ⋯ </DT>`
  - Used instead of the list item tag to identify the terms in a definition list.
  - `DT` stands for "Definition Term."
  - Ending tag is usually omitted.

- `<EM> ⋯ </EM>`
  - Semantic based style type.
  - Used to emphasize a portion of text.

- `<FONT> ⋯ </FONT>`
  - Used for changing the font specifications for any piece of text.
  - Attributes: `COLOR`, `FACE`, and `SIZE`.

- `<HEAD> ⋯ </HEAD>`
  - The first part of every HTML document.
  - Includes such items as the title of a page.

- `<Hi> ⋯ </Hi>`
  - Used to specify a heading.
  - i can be any value from 1 to 6.
  - Attribute: `ALIGN`.

- `<HR>`
  - No associated ending tag.
  - Produces a horizontal line or "rule."
  - Attributes: `ALIGN`, `NOSHADE`, `SIZE`, and `WIDTH`.

- `<HTML> ⋯ </HTML>`
  - Surrounds all text in an HTML document.

- `<I> ⋯ </I>`
  - Syntactic based style type.
  - Used to italicize a portion of text.

- `<IMG>`
  - No associated ending tag.
  - Used for including in-line images in HTML documents.
  - Attributes: `ALIGN`, `ALT`, `BORDER`, `HEIGHT`, `HSPACE`, `LOWSRC`, `SRC`, `USEMAP`, `VSPACE`, and `WIDTH`.

- `<KBD>` ⋯ `</KBD>`
  - Semantic based style type.
  - Used for describing keyboard input.

- `<LI>` ⋯ `</LI>`
  - Used to identify each item in an ordered or unordered list.
  - Ending tag is usually omitted.
  - Attributes: `TYPE` and `VALUE`.

- `<META>`
  - No associated ending tag.
  - Used to create client pull documents.
  - Attributes: `CONTENT`, `HTTP-EQUIV`, `NAME`, and `URL`.

- `<OL>` ⋯ `</OL>`
  - Used to specify an "Ordered List."
  - Attributes: `START` and `TYPE`.

- `<P>` ⋯ `</P>`
  - Used to split text into paragraphs.
  - Attribute: `ALIGN`.

- `<SMALL>` ⋯ `</SMALL>`
  - Semantic based style type.
  - Used to reduce the relative font size of text.
  - Produces the opposite effect of the big tag.

- `<STRIKE>` ⋯ `</STRIKE>`
  - Syntactic based style type.
  - Used to produce a line through the text to achieve a cross-out effect.

- `<STRONG>` ⋯ `</STRONG>`
  - Semantic based style type.
  - Used to indicate an even higher degree of emphasis than the emphasis tag.

- `<SUB> ··· </SUB>`
  - Syntactic based style type.
  - Used to generate a subscript.
- `<SUP> ··· </SUP>`
  - Syntactic based style type.
  - Used to generate a superscript.
- `<TABLE> ··· </TABLE>`
  - Used to define a table.
  - Useful for controlling the layout of information.
  - Attributes: `ALIGN, BGCOLOR, BORDER, CELLPADDING, CELLSPACING, HSPACE, VALIGN, VSPACE,` and `WIDTH`.
- `<TD> ··· </TD>`
  - Used to specify entries in the row of a table.
  - `TD` stands for "Table Data."
  - Ending tag is usually omitted.
  - Used within the table row tag of the table tag.
  - Attributes: `ALIGN, BGCOLOR, COLSPAN, NOWRAP, ROWSPAN,` and `WIDTH`.
- `<TH> ··· </TH>`
  - Used to specify table headings.
  - `TH` stands for "Table Header."
  - Ending tag is usually omitted.
  - Used within the table tag.
  - Attributes: `ALIGN, BGCOLOR, COLSPAN, NOWRAP, ROWSPAN,` and `WIDTH`.
- `<TITLE> ··· </TITLE>`
  - Used to specify the title of an HTML document.
- `<TR> ··· </TR>`
  - Used to position all items in the row of a table.
  - Used within the table tag.
  - `TR` stands for "Table Row."
  - Ending tag is usually omitted.
  - Attributes: `ALIGN, BGCOLOR,` and `VALIGN`.
- `<TT> ··· </TT>`
  - Syntactic based style type.
  - Used for displaying a portion of text in typewriter font.

- `<U> ... </U>`
  - Syntactic based style type.
  - Used to underline text.
  - Use sparingly, since some confusion can result as to whether or not the underlined item is a hyperlink.
- `<UL> ... </UL>`
  - Used to specify an "Unordered List."
  - Attribute: `TYPE`.
- `<VAR> ... </VAR>`
  - Semantic based style type.
  - Used to indicate a variable.

My URLs

For each chapter, we provide space for you to record your own URLs, as well as URLs associated with your Internet class. Many up-to-date references can be found in this book's accompanying Web presentations.

MISCELLANEOUS NOTES

My ISP's help information telephone number is:

_____

_____

_____

_____

_____

_____

_____

_____

_____

_____

_____

_____

_____

http://www.ohio.edu/information/chicago/index.html

THE INTERNET

Glossary

This glossary contains a list of important terms used in the text. Both the terms and their meanings are presented. Page numbers on which each concept is used can usually be traced through the index. This list is not intended to be exhaustive. For example, many of the HTML concepts are not included, but can easily be traced through the index.

account name—A name that identifies you to a computer; also called a userid.

alert box—A pop-up dialog box that contains an important message.

algorithm—A well-defined set of rules for carrying out a procedure.

alias—An easy-to-remember name associated with an e-mail address. The alias is saved by your e-mail software. Aliases can also be used in some operating systems to rename commands.

all-in-one search engine—See **metasearch engine**.

animated GIF—A series of GIF images that are displayed in rapid succession, thereby creating a simple animation.

anonymous file transfer—A mechanism that allows any user to transfer a file from a system.

applet—A small Java program. Applet means "little application."

Archie—A program that is useful for searching file archives.

article—A message that is sent to a newsgroup. Posting is a synonym for article.

attribute—A property associated with an HTML tag. For example, the anchor tag has an HREF attribute that is used to specify a hyperlink reference.

audio—Sound.

authenticate—A term meaning to verify that you are who you say you are.

baud rate or **bit rate**—A measure of the rate at which data is transferred over telecommunications lines. Many modems have a data transfer rate of 28.8 Kbps, that is, 28.8 thousand bits per second, or 28,000 baud.

binary transfer mode—A file transfer mode setting that allows binary files, such as executable programs and images, to be transferred.

bit rate—See **baud rate**.

blind carbon copy—A copy of an e-mail message that is sent to another user, but without the address for that user being displayed in the e-mail message's header.

bookmark—A way to identify a URL and make it easy to recall. The browser saves the URL and its title. The saved URL is also referred to as a bookmark.

Boolean query—Queries that involve the Boolean operations AND, OR, and NOT.

browser—A software application that provides an interface between users and the Internet. Netscape's Navigator and Microsoft's Internet Explorer are two popular browsers. Browsers are also called Web clients.

buffer—A location where data can be temporarily stored.

byte —A computer measurement equal to eight bits and used to indicate file size. It is also used in conjunction with time to indicate transfer rates.

cache—Local memory where copies of frequently used or recently requested Web pages are saved.

Caesar ciphers—A class of simple encryption schemes in which letters of the alphabet are rotated in a circular fashion. ROT13 is an example of a Caesar cipher.

Cascading Style Sheets—A method of specifying content that is common to a series of Web pages.

CGI script—See **script**.

chat room—An on-line forum where you can discuss your favorite subject interactively with other people located anywhere on the Internet.

ciphertext—An encrypted message.

circuit switching—A method of data transmission that was popular prior to packet switching.

clickable text—A hyperlink that consists solely of text.

client pull—A model of a dynamic document in which the browser (or client) initiates the document's change. This can be used to cause a document to "refresh" itself or to load a completely new document, among other things.

client-server model—The scheme in which many clients make requests to a small number of servers. The servers respond to clients' requests.

clip art—A collection of images that have been developed using image editing tools.

closed list—A mailing list for which the list owner either accepts or rejects subscriptions. It is not possible to subscribe automatically to a closed list. List membership is filtered and only postings from subscribers are accepted. Private list is a synonym for closed list.

collaborative computing—A computing platform that allows the sharing of information and resources among two or more users. Lotus Notes, Novell's Groupwise, and Microsoft Exchange are some applications that support collaborative computing.

color palette—See **palette**.

comment—A note, placed in an HTML file, that is not interpreted (or displayed) when the file is processed.

Common Gateway Interface (CGI)—A set of rules that specify how parameters are passed from programs to Web servers.

composite number—A number containing factors other than 1 and itself.

compression algorithm—Any computer algorithm that is used to reduce the size of files. On rare occasions, compression algorithms may increase the size of a file.

Computer Emergency Response Team (CERT)—An organization that sends out information about known security holes in software.

computer literacy—A basic knowledge of computer usage.

cookie—A piece of information saved by your Web browser to a file on your disk. The information can be retrieved by a Web server that your browser accesses.

copyright—A set of legal rights extended to an individual or company that has produced a creative work.

cross-post—The process of simultaneously submitting the same article to two or more newsgroups.

cryptographic algorithm—An algorithm that is used for either encoding or decoding information.

cryptography—The science of encoding and decoding information.

cyberspace—A popular term for the Internet.

default password—The initial password you are assigned to grant you access to an on-line item (for example, your computer account). Default passwords should be changed during your first access. The word default applies in other settings, usually with similar meaning.

deprecated tag—An HTML tag that should no longer be used because it is being phased out of the language.

digest—A collection of related articles, usually edited, that is posted as a single article to a newsgroup.

digested list—A mailing list in which postings are grouped by topic and sent out as batches instead of individually.

digital signature—A mechanism that can be used to sign an electronic document officially.

distance learning—Any form of teaching in which the instructor(s) and students are not located in the same room.

dithering—A process that approximates the color of each pixel in an image by using a combination of colors in a limited color palette.

Doctor HTML—A popular HTML syntax checker.

document area—The part of the browser window that is used to display the currently loaded document.

domain name system (DNS)—A distributed naming scheme in which unique names are assigned to computers on the Internet.

double key cryptography—See **public key cryptography**.

dynamic document—A document that has the capacity to change, either by client pull, server push, or some other mechanism.

dynamic IP address—The address assigned by your ISP when you connect to the Internet; this address usually changes each time you log on.

edited list—A mailing list in which items posted may be edited by the list owner or moderator.

e-mail—Messages that are sent electronically over a network. The term "e-mail" stands for "electronic mail."

e-mail address—An address that identifies a specific user's electronic mailbox, and has the form:

```
username@hostname.subdomain.domain
```

emoticons—Symbols made up of keyboard characters designed to express emotion; most commonly used in text-only communication, such as e-mail.

encryption scheme—A method of encoding information to make it secure.

event—An action or occurrence, usually initiated by a user.

event handler—A computer program that is executed in response to an event.

expired news—Old news that has been removed from a system.

eXtensible Markup Language (XML)—A Web page design language that will support user-defined tags.

e-zine—An electronic magazine.

file compression—A means of reducing a file's size by encoding the contents so that the file takes up less space.

file transfer—A way of transferring files from one computer to another computer, using a network.

firewall—A security mechanism that organizations use to protect their intranet from the Internet.

flame—A nasty, electronic response from an offended party.

flame war—A series of nasty, electronic responses that are part of the same newsgroup thread.

follow-up—A newsgroup article posted in response to a previous newsgroup article.

footer—The content displayed at the bottom of a Web page.

frame—An HTML feature that allows you to divide a browser's window into several independent parts.

freeware—Software that you can use at no charge. The author usually retains any copyright on it, and freeware frequently is unregistered. The source code is usually not provided.

frequently asked questions (FAQs)—Questions that many computer users ask. Because the answers to such questions are important to many people, they are usually collected and posted to either a mailing list or a newsgroup, or displayed on a Web page.

gigabyte—A billion bytes.

gopher—A menu-based Internet browsing tool that was very popular in the early 1990s.

graphical user interface (GUI)—A mouse-driven, rather than a keyboard-driven, graphically oriented computer interface.

groupware—The software that comprises a collaborative computing platform.

guest book—A mechanism that provides a way for readers visiting your Web pages to "sign-in" and leave you a note.

hash or **message digest**—The value computed by a hashing algorithm.

hashing algorithm—An algorithm that takes a plaintext message as input and then computes a value based on the message.

header—The content displayed at the top of a Web page.

helper or **helper application**—A stand-alone program that is used to process or display data that a Web browser is not able to handle.

helper application—see **helper**.

hit—A URL that a search engine returns in response to a query. Match is a synonym.

homepage—The Web page that is loaded when a browser is first activated; also, the first page in a set of related Web pages.

hot buttons—Single-click buttons in a browser that provide a number of convenient features.

HTML—HyperText Markup Language; the programming language in which most Web pages are written.

HTML converter—A program that takes one type of document as input and produces the same information in an HTML format.

HTML editor—A software editing tool that helps in developing HTML code.

HTML syntax checker—A program that processes an HTML document to see if there are any coding errors in it.

hyperlink—Text and/or graphics on a Web page that, when selected, will cause the browser to retrieve and render another Web page or graphic.

hypermedia—A Web document that contains any combination of audio, graphics, movies, or video (versus a document containing only text), as well as links, and navigational tools.

hypertext—Web pages that have hyperlinks to other pages or to other places on the same page.

hypertext transfer protocol (HTTP)—The rules that govern how hypertext is exchanged over the Internet.

image map—An image used in an HTML document with clickable areas that cause the loading of other documents.

in-line image—An image that is displayed as an HTML document loads.

Information Superhighway—A popular name for the Internet.

interlaced GIF—A form of GIF image in which the "whole image" starts to load initially, but appears blurry, and then comes into sharper focus as the download advances, until the image is finally complete.

Internet—A global system of networked computers, including their users and data.

Internet address—Numerical computer names that uniquely identify each computer on the Internet. Each address consists of four bytes, and each byte represents a decimal number from 0 to 255. The address is often represented by four decimal numbers separated by dots.

Internet Engineering Task Force (IETF)—A group that provides an open forum to facilitate communication between individuals dealing with matters related to the Internet.

Internet Explorer—The name of Microsoft's Web browser.

Internet Protocol (IP)—One of the primary protocols in the TCP/IP suite; IP specifies how data is routed from computer to computer on the Internet.

IP address—See **Internet address**.

Internet Protocol Version 6 (IPv6)—The latest version of the Internet Protocol (IP).

intranet—A private network in which access is limited to authorized users and a security measure known as a firewall is employed to prevent unauthorized users from gaining access.

Java—An object-oriented programming language that was developed by Sun Microsystems and is widely used to create dynamic Web pages.

Java-enabled—A browser that can run Java code.

JavaScript—A scripting language that is embedded in HTML and is useful for adding dynamic features to Web pages.

JavaScript-enabled—A browser that can run JavaScript code.

Java Virtual Machine—A computer program that allows you to run a Java application on a particular type of computer (for example, a Mac or a PC).

kill files—A filter based on names or topics you specify, for the purpose of blocking those newsgroup messages matching your criteria.

kilobit—1,000 bits; a unit of measurement often associated with modem transfer rates. Example: 28.8 Kbps (kilobits per second).

kilobyte—1,000 bytes; a unit of measurement often associated with file size or transfer rates (when combined with time).

list owner—A person in charge of a mailing list. Synonyms are list administrator, list coordinator, and list manager.

LISTPROC—A popular mailing list server program.

LISTSERV—A popular mailing list server program.

Local Area Network (LAN)—A privately owned computer network that is usually confined to a single building.

location area—The place in a browser window where URLs are entered and displayed.

lossless compression—A form of image compression in which no information is lost.

lossy compression—A form of image compression in which information is removed. The key is to delete information that has little or no impact on the appearance of the image.

lurker—A person who has subscribed to and reads a mailing list, but does not post messages. Such a person is said to lurk.

Lynx—The most popular text-based Web browser.

mailbox—A file that holds a user's e-mail messages.

mailer —A program that is used to compose, manipulate, and send e-mail. Synonyms are mail application, mail client, and mail program.

mailing list—A group of users with a shared interest, whose e-mail addresses are kept in an electronic list that can be used to send e-mail to each member on the list.

Majordomo—A popular mailing list server program.

megabyte—A million bytes; a unit of measurement often associated with file size or transfer rates (when combined with time).

menu bar—The place in a browser window where the headings of the main pull-down command menus are displayed.

message digest—See **hash**.

metasearch engine or **all-in-one search engine**—A search tool that calls on more than one other search engine to do the actual searching.

Metropolitan Area Network (MAN)—A computer network that spans an area about the size of a city. Such a network is larger than a LAN but smaller than a WAN.

mirror site—A site that contains a duplicate copy of a Web presentation from another site in order to reduce server traffic.

Multipurpose Internet Mail Extensions (MIME)—A system that is used by mailers and Web browsers to identify file contents by file extensions.

moderated newsgroup—A newsgroup that has a moderator.

Mosaic—The first widely popular graphical Web browser. It was developed by Marc Andreessen and several other graduate students at the University of Illinois in 1993.

multimedia—More than one type of media; any combination of two or more of animation, audio, graphics, text, and video.

Multi-User Dungeon or **Dimension (MUD)**—A real-time interactive game that takes place in an imaginary environment where multiple computer users can play simultaneously.

navigational tools—The buttons, hyperlinks, and images that allow a user to navigate a Web presentation.

netiquette—Informal rules of network etiquette.

Netscape—The Netscape Communications Company's Web browser.

Network News Transfer Protocol (NNTP)—The protocol that is used for distributing news articles.

newbie—A person who only recently joined a mailing list.

news administrator—A person who is in charge of running a news server.

newsfeed—A news server that provides recent articles to a news client. The term also encompasses the process of delivering the news articles.

newsgroup—An on-line forum that allows users from all over the world to participate in a discussion about a specific topic.

news moderator—A person associated with a specific newsgroup who reads and critically evaluates all articles submitted for posting to the newsgroup and then decides which articles should be posted.

newsreader—A program that serves as the user interface for newsgroups and allows a user to read, post, subscribe, and unsubscribe from the newsgroup. News client is a synonym.

news server—A computer that saves, forwards, and manages news articles. Normally each organization runs its own news server and limits access to just its customers or user community.

object-oriented programming (OOP)—A computer programming style that uses classes and methods.

open list—A mailing list to which anyone may subscribe. Such a list typically accepts posts even from users that are not subscribers.

packet—A small piece of a message that is transported over the Internet.

packet switching—The technology employed to route messages over the Internet.

page—See **Web page**.

pages—See **Web pages**.

palette or **color palette**—A defined group of distinct colors available for a particular purpose or use.

password—A secret code you provide when you log in that is used to authenticate you to a computer.

pattern matching query—A query formulated using a keyword or a group of key-words.

plaintext—A message in its original form; that is, not encoded.

plug-in—A software application designed to extend the functionality of a Web browser. Plug-ins are launched from within the browser and are capable of playing audio, showing movies, and running animations, among other things.

Point-to-Point Protocol (PPP)—A protocol that is widely used for transferring pack-ets over telephone lines.

post—The process of sending an article to a newsgroup; also, an article that is sent to a newsgroup.

Pretty Good Privacy (PGP)—An e-mail security package developed by Phil Zim-mermann. It includes authentication, compression, digital signature, and privacy capabilities, and uses the RSA encryption algorithm.

prime number—A number whose only factors are 1 and itself.

private key cryptography—An encryption scheme in which both the sender and the receiver share the same private key.

Project Gutenberg—An on-line book project whose goal is to put 10,000 books on the Web by the end of the year 2001.

protocol—A set of precisely specified rules for carrying out a procedure.

prototyping—The process of designing a system to work out the design deficiencies before building the final product.

public domain software—Free software available over the Internet; the source code is seldom available, but there may be guidelines as to how you are "allowed" to modify the original source code.

public key cryptography—An encryption scheme in which a message encrypted using a private key can only be decrypted using its matching public key.

query—Information entered into a form on a Web page, describing the topic on which information is sought. A query is usually not phrased as a question.

query semantics—A set of rules that defines the meaning of a query.

query syntax—A set of rules describing what constitutes a legal query. On some search engines, special symbols may be used in a query.

recursive algorithm—An algorithm that calls itself.

register (a Web page)—The process of submitting the URL of a Web page to a search engine or directory.

relevancy score—A value that indicates the closeness with which a URL matches a query. It is usually expressed as a value from 1 to 100, with the higher score meaning more relevant.

remote login—A method of logging into another (distant) computer from the one to which you are currently connected. Once logged in, you are able to execute commands on the remote computer.

render—The browser process displaying a Web page on the screen.

Request for Comments (RFC)—The official documents provided and distributed by the Internet Engineering Task Force.

result set—The list of hits returned by a search or metasearch engine.

resolver—A program that translates between domain names and IP addresses.

revolving advertisements—Advertisements, displayed on Web pages, that change every time you visit the page, or while you are viewing the page.

RGB color model—A way to represent colors as combinations of red, green, and blue.

Rotation 13 (ROT13)—A simple encryption scheme in which letters are rotated 13 positions further down the alphabet.

router—A special-purpose computer that directs packets of data along a network.

RSA encryption scheme—The most widely used public key encryption scheme. It is named for its developers, Rivest, Shamir, and Adleman.

scanner—A device that converts images to a digital format.

script or **CGI script**—Any program that is run by a Web server in response to a user's request.

scroll bar—Arrows along the side or bottom of a window that allow the user to display a different part of a document that is larger than the screen size.

search engine—A search tool that allows a user to enter queries. The program responds with a list of matches from its database. A relevancy score for each match and a clickable URL are usually returned.

search tool—Any mechanism for locating information on the Web; usually refers to a search or metasearch engine, or to a directory.

secure document—An electronic file that has been encoded so only those who know how to decode the file can read the information it contains.

secure server—A server that sends and receives encrypted (secure) messages.

semantics—The meaning associated with commands or statements in a given computer language; the interpretation of the syntax of a computer language.

semantic based style type—An HTML style that pertains to meaning; for example, emphasis and citation.

Serial Line Internet Protocol (SLIP)—A protocol that is used for transferring packets over telephone lines.

server—A computer that satisfies user (client) requests.

server push—A model of a dynamic document in which the server initiates the document's change. Server push is not accomplished using HTML tags.

server-side include (SSI)—A way to send a command to a Web server from inside an HTML document.

7 by 24 machine—A machine that runs 7 days a week, 24 hours a day.

shareware—Software that you can download and test out for a brief trial period. If you decide to use the software, you pay a small fee. Many times, the fee is collected on an honor system basis.

signature file—A file that contains an e-mail signature. A person's signature file is usually appended to all e-mail messages they send.

s-mail—See **snail mail**.

smiley—A happy face, written as : -); an example of an emoticon.

snail mail—Regular postal mail; also referred to as **s-mail**.

sniffing—The process of tapping into a network and reading the packets that are being transmitted.

spam—Inappropriate or junk e-mail.

static IP address—A permanently assigned IP address.

status bar—A field the browser uses to convey helpful (and current) information to the user.

stemming—The process a search tool uses to add variations to the endings of words you query on, to turn up more hits.

storyboard—A sketch of how a browser's window is to be partitioned when a document that uses frames is being designed.

streaming—The process of buffering data and using it to achieve a continuous-play effect, while the next part of the data is being transported in parallel over the Internet; widely used in the context of multimedia.

subscribe—The procedure you follow to join a newsgroup. Once subscribed you will receive the new postings of the newsgroup. The subscription is also used to track which messages in a group you have seen. In most cases, subscriptions do not cost money.

surfer—A person who spends time exploring the Web.

syntax—The rules or structure that describe the form of statements in a computer language.

syntactic based style type—An HTML style that pertains to form, not meaning; for example, boldface and italics.

tag—The name given to HTML commands. For example, the image tag, `<IMG>`, is used to include an image in an HTML document. Tags usually come in matched pairs, such as `<FORM>` and `</FORM>`.

target—The location (frame or window) to which a hyperlink is directed.

Telnet—A program that allows you to log into a remote computer.

thread—A collection of one or more follow-up articles, together with the original posting in a newsgroup.

thumbnail sketch—A reduced-size image that is used to give a reader a preview of an image, so they can decide whether or not they would like to spend the time loading the full image.

tiling or **tiling algorithm**—A method used to fill in a background on a Web page by taking a small image and laying out repeated copies of that image until it covers the browser's entire window.

title bar—The location in a browser window where the HTML document title is displayed.

toolbar—The area in a browser window for accessing a number of single-mouse-click commands.

tooltip—Usually a light-colored, dialog box that displays helpful information when you mouse over an item in the browser window.

Transmission Control Protocol (TCP)—One of the primary protocols in the TCP/IP suite. TCP defines a set of rules for allowing computers on the Internet to communicate.

Transmission Control Protocol/Internet Protocol (TCP/IP)—The protocol suite that determines how computers connect, send, and receive information on the Internet.

transparent GIF—A GIF image that creates a visual effect in which the image appears to be "floating" on a Web page.

triage—A strategy designed to process the most important items first, such as dealing with priority e-mail messages first and less important messages second.

Trojan horse—A program within which code is hidden. When the hidden code is triggered, it might release a virus, permit unauthorized access to the computer, or destroy files and data.

under construction—A phrase used to described unfinished Web pages.

Uniform Resource Locator (URL)—A Web page address, such as

`http://www.playground.com/~killface/cats.html`

universal service—Any service that is available worldwide. For example, many people consider the telephone to be a universal service.

UNIX—A widely used computer operating system, particularly in academic and research environments.

unmoderated newsgroup—A newsgroup that has no moderator.

unsubscribe—The process of removing your name from a newsgroup to which you have previously subscribed.

userid—A name that identifies you to a computer; also called a "user name" or "account name."

vacation program—A program that can be set up to reply automatically to each e-mail message you receive. Such a program is usually installed when you are going to be away for a week or more.

videoconferencing—A system designed to permit real-time interaction between multiple parties. It can involve one or more of the following: real-time talk or chat, whiteboard graphics, audio, black and white video, or color video.

virtual reality—A three-dimensional simulation of a real or imagined environment, using computers.

virus—A program that, when run, is able to replicate and embed itself within another program, usually with the intent of doing damage.

way-station—A news server that functions as a newsfeed for at least two (and usually many more) other sites.

Web—See **World Wide Web**.

Web directory—A hierarchical representation of hyperlinks to Web presentations.

Webmaster—A person who maintains, creates, and manages a Web presentation, and is responsible for responding to questions and comments. The term Webmaster usually implies a certain minimal level of knowledge. Web manager is a synonym.

Web page or **page**—A file that can be read over the World Wide Web.

Web pages or **pages**—The global collection of documents associated with and accessible via the World Wide Web.

Web presentation—A collection of associated and hyperlinked Web pages that usually has some underlying theme.

Web server—A computer that satisfies requests for Web pages.

Web site—An Internet entity that publishes Web pages. A Web site typically has a computer serving Web pages, whereas a Web presentation is the actual Web pages themselves.

white pages—A database that serves as an on-line telephone book.

Wide Area Information Service (WAIS)—A database search system that employs sophisticated feedback mechanisms. Its popularity has been waning in recent years.

Wide Area Network (WAN)—A computer network that spans a large geographical area, such as a country or a number of cities.

World Wide Web or **Web**—An application that utilizes the Internet to transport hypertext/multimedia documents. Synonyms are WWW, W^3, and W3.

World Wide Web Consortium (W3C)—A group that provides an open forum to facilitate communication between individuals dealing with matters related to the World Wide Web.

World Wide Wait Problem—The delay experienced on the Internet, caused by the tremendous popularity of the Web.

worm—A stand-alone program that tries to gain access to computer systems via networks.

Yahoo!—A popular search engine and directory developed by former Stanford graduate students David Filo and Jerry Yang.

Bibliography

Notes

This book deals with a rapidly emerging area. Most of our research was conducted on-line. As a result, many of our references are Web presentations. Each citation for a Web presentation provides its title (as capitalized by the author), date, and URL. For the sake of readability, we have omitted the leading `http://` and also any trailing `/` on URLs. Sometimes we just include the URL for the main page of a presentation, although we may have read through many of its subpages. In this bibliography, URLs are sorted by title, and other references are sorted by the author's last name. We have included some extra references for more advanced topics.

Most presentations do not provide dates indicating when the document was written, nor are the original document creation dates available on-line (or anywhere, for that matter). Some documents written in 1996 may have been completely overhauled in 1997, meriting a 1997 date, whereas others may have only been modified slightly, meriting a 1996 designation. Some authors include "last update" messages in headers, while others do not. Due to the inherent inconsistencies in dating Web presentations, the date we used in most cases indicates when we successfully accessed the site. Most of the Web presentations were written in 1997. We would appreciate any corrections to the dates from the authors of the presentations and also notices of changed URLs.

The Web presentation that accompanies this book contains a number of useful URLs, sorted by chapter, plus many other additional hyperlinks. Some URLs are from these references, while others are new.

1   AmericaNet.Com Since July 4, 1995, 1997. `www.americanet.com`

2   Apple QuickTime VR Home, 1997. `qtvr.quicktime.apple.com`

3   AprilFools.com, 1997. `www.aprilfools.com/home.htm`

4   Assessing Your Scanner Options in a Buyer's Market, 1997. `www.zdnet.com/cshopper/content/9611`

5   Atlantic Records, 1997. `www.atlanticrecords.com`

6   10 Big Myths About Copyright Explained, 1997. `www.clari.net/brad/copymyths.html`

7   Bridging the Gap Between Theory and Practice—the Journal of Electronic Publishing, 1997. `www.press.umich.edu/jep/03-01/JEA.html`

8   A Brief History of the Internet, 1997. `www.isoc.org/internet-history/#Introduction`

9   A Brief History of Type, 1997. `tug.cs.umb.edu/tetex/html/fontfaq/cf_28.html`

10   A brief intro to copyright, 1997. `www.clari.net/brad/copyright.html`

11   BUILDER.COM—Web Authoring—20 questions about HTML 4.0, 1997. `www.cnet.com/Content/Builder/Authoring/Html40/?bl.auth.5`

12   Choosing an ISP, 1997. `www.currents.net/resources/netprov/intquest.html`

13   CNET reviews—comparative reviews—11 HTML editors, 1997. `www.cnet.com/Content/Reviews/Compare/11htmleds`

14   Collaborative Networked Communication: MUDs as Systems Tools, 1997. `www.ccs.neu.edu/home/remy/documents/cncmast.html`

15   Comer, Douglas. *The Internet Book*. 2nd ed. Upper Saddle River, NJ: Prentice-Hall, Inc., 1997.

16   Computers and Copyrights: Bibliography, 1997. `www.iat.unc.edu/guides/irg-04.html`

17   Cookie Central, 1997. `www.cookiecentral.com`

18   cookie—PC Webopaedia Definition and Links, 1997. `www.sandybay.com/pc-web/cookie.htm`

19   Copyright and Fair Use in the Digital Age: Q&A with Peter Lyman, 1997. `www.educom.edu/web/pubs/review/reviewArticles/30132.html`

20   Copyright Clearance Center Online, 1997. `www.copyright.com`

21   Copyright in a digital age, 1997. `www.onlineinc.com/articles/onlinemag/weiner975.html`

22   Copyright in the New World of Electronic Publishing, 1997. `www.press.umich.edu/jep/works/strong.copyright.html`

23   The Copyright Question — Internet World January 1997, 1997. `www.internet.com/search`

24   Cornell University's CU-SeeMe Page, 1997. `cu-seeme.cornell.edu`

25   Cunningham, Sally Jo. Teaching students to critically evaluate the quality of internet research resources. *SIGCSE Bulletin*, 29(2), June 1997, pp. 31–34.

26   Debunking Myths About Internet Commerce, 1997. `commerce.ssb.rochester.edu/papers/comment.htm`

27   Deja News, 1997. `www.dejanews.com`

28   The Digital Press, 1997. `www.iw.com/1995/09/feat58.htm`

29   Electronic Publishing, 1997. `www.parkcce.org/index.html`

30   Emerging Technologies, 1997. `commerce.ssb.rochester.edu/emerging.html`

31   Fakemail, pranks, & gags: Rubberchicken.com, 1997. `www.rubberchicken.com`

32   Federal-Express Sets New Strategic Direction With On-line Ordering, 1997. `www.informedusa.com/t/fedexautoorder10.11.html`

33   FedEx: Electronic Commerce Connections, 1997.
     `www0.fedex.com/connections`

34   Finding An ISP That You'll Really Enjoy Using, 1997.
     `www.accesss.digex.net/~mccork/isp.html#intro`

35   FNC Resolution: Definition of Internet, 1997.
     `www.fnc.gov/Internet_res.html`

36   food-online.com—Corporate Background, 1997.
     `www.food-online.com/pr/corp_bg.htm`

37   About a Framework for Global Electronic Commerce, 1997.
     `www.whitehouse.gov/WH/New/Commerce/about-plain.html`

38   A Framework for Global Electronic Commerce Executive Summary, 1997.
     `www.whitehouse.gov/WH/New/Commerce/summary-plain.html`

39   Frequently Asked Questions: Basic Information about MUDs and MUDding,
     1997.   `www.cs.okstate.edu/~jds/mudfaq-p1.html`

40   Frequently Asked Questions: MUD Clients and Servers, 1997.
     `www.cs.okstate.edu/~jds/mudfaq-p2.html`

41   The Future of Electronic Journals—Neurotrophism, 1997.
     `www.physiol.washington.edu/ngf/editorial.HTM`

42   GIF Animations, 1997.   `www.babylon6.demon.co.uk/gif.html`

43   DiP Pixel Explorations—A GIF Transparency FAQ, 1997.
     `www.bruin.ucla.edu/TaskForce/transparent←`
     `/gif_faq.htm#PSPandGIFcon`

44   Gilster, Paul. *The New Internet Navigator*. New York, NY: Wiley and Sons, Inc.,
     1997.

45   Graphics FAQ, 1997.   `www.aldridge.com/faq_gra.html`

46   Early Adopters: Groceries To Go, 1997.
     `www.food-online.com/pr/gtg.htm`

47   Groceries To Go—FAQ, 1997.
     `www.food-online.com/gtg/gtgfaq1.htm#1`

48   Groupware Grows Up, 1997.   `techweb.cmp.com/iw/569/69iugrp.htm`

49   Hafner, Katie, and Matthew Lyon. *Where Wizards Stay Up Late: The Origins of the
     Internet*. New York, NY: Simon and Schuster, 1996.

50   Hahn, Harley. *The Internet: Complete Reference*. 2nd ed. Berkeley, CA: Osborne
     McGraw-Hill, 1996.

51   The History and Characteristics of Zines, 1997.
     `www.thetransom.com/chip/zines/resource.html`

52   The History of the Internet, 1997.
     `www.davesite.com/webstation/net-history.shtml`

53   How An ISP Really Works, 1997.
     `www.mindspring.com/~mcgatney/ispwork.html`

54   How the Internet Works, 1997.   `www.iw.com/1996/howitworks.html`

55   How to Buy Flatbed Scanners, 1997.
     `www1.zdnet.com/complife/rev/9707/scanner1.html`

56   How To Select An Internet Service Provider, 1997.
     `web.cnam.fr/Network/Internet-access/how_to_select.html`

57   IBM Cryptolope Home, 1997.   `www.cryptolope.ibm.com/cryptolp.htm`

58   Imagemap Help Page—Instruction, 1997.   `www.ihip.com`

59   Inclusions in HTML documents, 1997.
     `www.w3.org/TR/WD-html40-970708/struct/includes.html#h-7.7.1`

60   Inline Images Frequently Asked Questions, 1997.
     `galway.informatik.uni-kl.de/./comp/Mosaic/inline-images.html`

61   Intelligent Manufacturing—December 96: Collaborative Computing, 1997.
     `www.lionhrtpub.com/IM/IM-12-96/autofact.html`

62   The Internet Index Home Page, 1997.   `www.openmarket.com/intindex`

63   Internetter: SSI—The WebMasters Secret Weapon, 1997.
     `www.internetter.com/papers/ssi.html`

64   InterTrust Industry Applications, 1997.
     `www.intertrust.com/products/applications.html`

65   Introducing MUDs!, 1997.   `andes.ip.ucsb.edu/~krend/muds.html`

66   An Introduction to Electronic Data Interchange, 1997.
     `www.nlc-bnc.ca/publications/netnotes/notes6.htm`

67   IRC INTRO—Introduction to IRC for People using Windows, 1997.
     `www.mirc.co.uk/ircintro.html`

68   JPEG image compression FAQ, part 1/2, 1997.
     `www.cis.ohio-state.edu/hypertext/faq/usenet/jpeg-faq↩`
     `/part1/faq.html`

69   DiP presents the JPiG Project, 1997.
     `www.algonet.se/~dip/JPiG/jpig_1a.html`

70   Lemay, Laura. *Teach Yourself Web Publishing with HTML 3.2 in a Week.* Indianapolis, IN: Sams.net, 1996.

71   Life on the Internet: Net Timeline, 1997.   `www.pbs.org/internet/timeline`

72   Linux IPv6 FAQ/HOWTO, 1997.   `www.terra.net/ipv6`

73   Liszt, the mailing list directory, 1997.   `www.listz.com`

74   Lynx Users Guide Version 2.5, 1997.
     `tincan.tincan.org/help/lynx/lynx_help/Lynx_users_guide.html`

75   Mars Pathfinder in VRML!, 1997.
     `www.ncsa.uiuc.edu/mars/vrml/vrml.html`

76   Messaging Magazine—Collaborative Computing: Empowering People to Accelerate the Decision Making Process, 1997.
     `www.ema.org/html/pubs/mmv1n5/collabco.htm`

77   Messaging Magazine—Three Trends in Collaborative Computing, 1997.
     `www.ema.org/html/pubs/mmv1n5/3trends.htm`

78   Messaging Magazine—The New Business Paradigm, 1997.
     `www.ema.org/html/pubs/mmv1n5/paradigm.htm`

79   Messaging Magazine—Security in Collaborative Computing Environments, 1997.
     `www.ema.org/html/pubs/mmv1n5/securcc.htm`

80   Musciano, Chuck, and Bill Kennedy. *HTML: The Definitive Guide.* Sebastopol, CA: O'Reilly and Associates, Inc., 1996.

81   The Net and Netizens: The Impact the Net has on People's Lives, 1997.
     `www.columbia.edu/~rh120/ch106.x01`

82   NetscapeWorld—Cookies offload server overhead and can reduce some client overhead issues, 1997.

www.netscapeworld.com/netscapeworld/nw-07-1996↩
/nw-07-cookies.html

83   NetscapeWorld—Use Cookies to Analyze User Activity & Create Custom Web
     Pages — February, 1997.
     www.netscapeworld.com/netscapeworld/nw-02-1997↩
     /nw-02-cookiehowto.html

84   *The New York Times* on the Web, 1997.   www.nytimes.com

85   ONSALE, Inc., 1997.   www.onsale.com

86   Paulallen.com, 1997.   www.paulallen.com

87   *PC Magazine*: VRML Brings 3-D Worlds to the Web, 1997.
     www8.zdnet.com/pcmag/issues/1614/pcmg0060.htm#top

88   Platform for Internet Content Selection (PICS), 1997.   www.w3.org/PICS

89   Portable Network Graphics Home Page, 1997.   www.wco.com/~png

90   Professional Software Engineering MUD Description and Links Page, 1997.
     www.professional.org/pse/mud.html

91   Project Gutenberg—History and Philosophy, 1997.
     promo.net/pg/history.html

92   QTVR FAQ, 1997.   www.convrgence.com/smallsite/QTVRFAQ2.html

93   The Role of Publishers in the Digital Age—Educom Review, 1997.
     www.educom.edu/web/pubs/review/reviewArticles/30344.html

94   Royalties, Fair Use & Copyright in the Electronic Age—Educom Review, 1997.
     www.educom.edu/web/pubs/review/reviewArticles/30630.html

95   Scanning For the Rest of Us, 1997.
     www.macworld.com/pages/march.97/Feature.3297.html

96   A few scanning tips, 1997.   www.cyberramp.net/~fulton/scans.html

97   Scanning Tips, 1997.   www.jasc.com/scantip.html

98   Seltzer, Richard, Eric Ray, and Deborah Ray. *The AltaVista Search Revolution.*
     Berkeley, CA: Osborne McGraw-Hill, 1997.

99   The Short Little Guide To Selecting An ISP That You'll Enjoy, 1997.
     www.mindspring.com/~mcgatney/ispfast.html

100 SLIP/PPP Homepage, 1997.   sunsite.nus.sg/pub/slip-ppp

101 The slow evolution of electronic publishing, 1997.
     www.research.att.com/~amo

102 The Spider's Web: News Feature—How the HTTP Cookies Crumble, 1997.
     www.incontext.ca/spidweb/may15_96/news/cookie8.htm

103 Tanenbaum, Andrew. *Computer Networks.* Upper Saddle River, NJ: Prentice-Hall,
     Inc., 1996.

104 TechTools—ISP, 1997.   techweb2.web.cerf.net/tools/isp

105 TechWeb, 1997.
     www.techweb.com/se/directling.cgi?WIN1997090150120

106 The Economist: Internet, too cheap to meter & the World Wide Wait 19 Oct 96,
     1997.   www-uvi.eunet.fr/hacking/nov13=17nov96-6.html

107 the truth about cookies—Christopher Barr, 1997.
     www.cnet.com/Content/Voices/Barr/042996

108   Transcopyright: Dealing with the Dilemma of Digital Copyright—Educom
       Review, 1997.
           `www.educom.edu/web/pubs/review/reviewArticles/32132.html`

109   TWAIN White Paper, 1997.   `www.twain.org`

110   Videoconferencing FAQ, 1997.   `www.bitscout.com/faqtoc.htm`

111   Vint Cerf On the Past, Present & Future of All Things Internet, 1997.
           `www.wiredguru.com/cd2.html`

112   VRML Authoring & 3D Modeling Software, 1997.
           `www.webdeveloper.com/categories/vrml/vrml_editors.html`

113   comp.lang.vrml Frequently Asked Questions, 1997.
           `hiwaay.net/~crispen/vrml/faq.html#q1`

114   The Virtual Reality Modeling Language, 1997.
           `tom.di.uminho.pt/vrmlut/frmstrct.htm`

115   A VRML Primer, 1997.
           `www.ncsa.uiuc.edu/mars/vrml/primer/primer.html`

116   The VRML Repository, 1997.   `www.sdsc.edu/vrml`

117   VRML Update, 1997.   `www.meshmart.org/vrmlup.htm`

118   W3C Electronic Commerce Area, 1997.   `www.w3.org/Payments`

119   W3C Recommendations Reduce World Wide Wait, 1997.
           `www.w3.org/pub/WWW/Protocols/NL-PerfNote.html`

120   The Wall Street Journal Interactive Edition, 1997.
           `interactive6.wsj.com/home.html`

121   Web Graphics Format Page, 1997.   `www.jasc.com/filetip.html`

122   Weblynx—Step By Step Guide to VRML, 1997.   `weblynx.com.au/guide.htm`

123   The Web Multimedia Tour—Virtual Reality, 1997.
           `ftp.digital.com/webmm/vr.html`

124   Web Review: VRML — History, 1997.
           `webreview.com/sept29/features/vrml/history.html`

125   Welcome to Internet Travel Agency, Inc., 1997.
           `www.internettravelagency.com`

126   Welcome to the Ultimate Band List, 1997.   `ubl.com`

127   What is a MUD, actually?, 1997.
           `www.cwrl.utexas.edu/moo/mudhandouts/beginning.html`

128   What Is Project Gutenberg?—rec.art.books FAQ, 1997.
           `www.cis.ohio-state.edu/hypertext/faq/usenet/books/faq↩`
           `/faq-doc-6.html`

129   Whither the electronic journal?—SLS UK User Group, 1997.
           `www.lib.ic.ac.uk:8081/leah.htm`

130   Wilson, Lee. *The Copyright Guide*. New York, NY: Allworth Press, 1996.

131   With mars under its belt, VRML good to go, 1997.   `www.listz.com`

132   Yeager, Nancy, and Robert McGrath. *Web Server Technology*. San Francisco, CA:
       Morgan Kaufmann Publishers, Inc., 1996.

Index

A

A, 73
   HREF, 78
   NAME, 80
absolute URL, 57
account name, 299
Acrobat, 134
active hyperlink, 67
ADDRESS, 204
address book, 11
Adleman, Leonard, 308
Adobe Acrobat, 134
Advanced Research Projects Agency,
      91
advertisements, 112
AFAIK, 112
alert box, 299
algorithm, 12, 299
   compression, 301
   computer, 12
   hashing, 303
   RSA, 308
   tiling, 66, 310
alias, 10, 299
   local, 11
   private, 11
   public, 11
ALINK, 67
all-in-one search engine, 165, 299, 305
Allen, Paul, 137
ALT, 84
AltaVista, 115, 165, 166
America Online, 31, 251
American Standard Code for
      Information Interchange, 21
AmericaNet, 115
&, 225
Amphlett, Christina, 12
analog, 99
Andreessen, Marc, 94, 306
animated GIF, 299

anonymous file transfer, 196, 299
AOL, 251
   NetFind, 163, 165
Apache, 52
Appalachian Trail, 160
applet, 89, 145, 299
Archie, 198, 299
archives, 197
ARPA, 91
ARPANET, 91
   host, 91
arrow keys, 259
article, 299
ASCII, 21, 35, 257
   art, 22
Atlantic Records, 114
attribute, 47, 48, 299
AT&T, 92, 100
auction, 117
audio, 36, 123, 299
authenticate, 29, 299

B

B, 207
BACKGROUND, 65
background color, 64
backtrack, 182
bandwidth, 123
BASE, 57
   HREF, 58
   TARGET, 61
BASEFONT, 56
   SIZE, 56
baud rate, 99, 299, 300
BBN, 7, 91
The Beatles, 173
Because It's Time Network, 92
Bell Labs, 92
Berners-Lee, Tim, 93
BGCOLOR, 65

bibliographic record, 184
BIG, 207
Bigfoot, 166
biggie small, 12
binary
   digit, 62
   file, 195
   transfer mode, 195, 299
biography, 137
bit, 62
   rate, 99, 299, 300
BITNET, 92
blind carbon copy, 15, 300
BLINK, 210
blue ribbon, 110
BODY, 64, 211
   ALINK, 67
   BACKGROUND, 65
   BGCOLOR, 65
   LINK, 67
   TEXT, 65
   VLINK, 67
Bolt, Beranek, and Newman, 7, 91
bookmark, 40, 129, 130, 300
   list, 130
Boole, George, 172
Boolean
   algebra, 172
   query, 172, 300
bot, 180
bounce, 31
BR, 204, 244
breadth-first search, 181
brightness, 48
browser, 44, 129, 300
   definition, 37
buffer, 300
bullet, 219
bulletin board system, 251
business, 139
   advertising, 115
   marketing, 115

on-line, 115
partnership, 115
retail, 116
service, 116
software, 116
subscription, 116
busy signal, 108
button
    Compose, 23
    File, 23
    Reply, 23
byte, 13, 102, 300

C

cable television, 99
cache, 40, 129, 300
Caesar cipher, 300
camera, 123
CAPTION, 241
    ALIGN, 241
carbon copy, 15
Cascading Style Sheets, 106, 202, 300
case
    law, 110
    sensitive, 6, 48, 72, 277
CD, 111
CDA, 120
cells, 231
censorship, 111
CENTER, 217
centering, 217, 232
central mail spool, 29, 31
CERN, 93
CERT, 301
CGI, 301
    script, 300, 308
channel surfing, 114
chat room, 251, 300
child pornography, 110
cipher, Caesar, 300
ciphertext, 300
circuit switching, 91, 300
CISCO Systems, 96
CITE, 203
Clark, Jim, 94
Cleese, John, 4
clickable
    image, 78
    text, 78, 300
client, 100
    pull, 300
client-server model, 100, 300
Clinton, Bill, 120
clip art, 300
closed list, 300

closing tag, 49
CNET Search.com, 163
CODE, 206
collaborative computing, 87, 121, 300
COLOR, 68
color
    background, 64
    palette, 301, 307
comment, 70, 301
Common Gateway Interface, 301
Communications Decency Act, 120
Communicator, 39
compact disk, 111
compress, 196
compression, 196
    algorithm, 301
    lossless, 305
    lossy, 305
computer
    algorithm, 12
    literacy, 301
    network, 91, 95, 99
    virus, 35, 36, 92, 187, 193, 198
Computer Emergency Response
        Team, 301
Computer Science Network, 92
content style type, 202
converter, HTML, 245, 303
cookie, 118, 141, 301
    setting, 118
copper wire, 99
copyright, 71, 301
corrections, xvii
country code, 8
cranberry, 48
crawler, 180
credit card, 118
    electronically safe, 119
    risk, 119
    secure, 119
    security, 119
    transaction, 119
cryptography
    double key, 302
    private key, 307
    public key, 302, 307
CSNET, 92
CSS, 106
culture, 108
current directory, 276
CUSeeMe, 123
Cyber Patrol, 111
Cyber Sentry, 111
cyberspace, 111, 301

D

DARPA, 91
dash (-), 274
DD, 221
debugging
    HTML, 201
default, 301
    password, 13
Defense Advanced Research Projects
        Agency, 91
definition list, 217, 221
Dell Computer Corporation, 114
Denellio, Frank, 10
deprecated, 210
    tag, 301
depth, 155
depth-first search, 181, 182
design and code, 157
Desktop Videoconferencing, 123
dial up, *see* ISP
    busy signal, 252
    local number, 253
    problem, 256
    testing ISP, 255
digested list, 301
digital signature, 301
digitizer, 123
direct search, 165
directories, 162
    search, 161
directory, 162, 311
    hidden, 275
disclaimer, 169
distance learning, 301
distribution list, 11
    private, 11
dithering, 301
DIV, 217
    ALIGN, 217
divider, 229
Division of Motor Vehicles, 6
DL, 221
DNS, 8, 92
Doctor HTML, 301
document
    area, 38, 302
    creation, 47
    indexing, 184
documentation, 71
domain name, 102, 253
    country, 8
    server, 103
    space, 8, 102, 302
    U. S., 8
Domain Name System, 92
dot file, 275

double key cryptography, 302
DT, 221
DTVC, 123
dynamic
    document, 302
    IP address, 103, 302

E

e-mail, 1, 302
    address, 5, 302
    alias, 10
    attach, 21
    attachment, 16
    basics, 26
    Bcc, 15, 19
    bomb, 5
    bounce, 31
    business, 34
    Cc, 15, 19
    composition, 16
    Date, 15
    directory, 22
    filter, 34
    flame, 20
    folder, 22
    forwarding, 4, 25
    From, 14
    greeting, 15
    header, 14
    help, 19
    inbox, 27
    message body, 16
    MIME attachment, 16
    only service, 255
    programs, 2
    recommendations, 33
    reply, 18
    signature, 16
    spell check, 20
    Subject, 15, 17
    To, 15, 17
    vacation, 33, 311
    workings of, 26
e-zine, 302
edited list, 302
editor
    HTML, 303
    text
        emacs, 257
        Pico, 257
        vi, 257
electronic mail, *see* e-mail
elm, 2
EM, 203
emacs, 257

emoticon, 20, 111, 302
encapsulated postscript, 36
encryption scheme, 302
    RSA, 308
ending tag, 49
Environmental Defense Fund, 114
equipment, connect to Internet, 254
escape sequence, 190
Eudora, 2
evaluator, 178, 179
Evans, Janet, 174
event, 302
event-handler, 302
Excite, 163, 165
expired news, 302
eXtensible Markup Language, 302

F

FACE, 69
fan-out, 155
FAQ, 215, 303
    tables, 244
favorite, 130
Federal Express, 115
Federal Networking Council, 88
FedEx BusinessLink, 115, 116
Fetch, 195
fiber-optic, 99
file, 38
    archives, 197
    compression, 196, 302
    dot, 275
    extension
        aif, 36
        aifc, 36
        aiff, 36
        au, 36
        avi, 36
        eps, 36
        gif, 36, 83
        gz, 196
        htm, 36
        html, 36
        jpe, 36
        jpeg, 36
        jpg, 36, 83
        mid, 36
        midi, 36
        mov, 36
        movie, 36
        mpeg, 36
        mpg, 36
        png, 83
        ps, 36
        qt, 36

        ra, 36
        ram, 36
        roff, 36
        sgml, 36
        snd, 36
        t, 36
        tex, 36
        tiff, 36
        tr, 36
        txt, 36
        wav, 36
        wrl, 36
        Z, 196
        zip, 196
    hidden, 275
    insertion, 22
    permission, 49
    protection, 49
    transfer, 3, 89, 187, 193, 302
        graphical, 193
File Transfer Protocol, 195
Filo, David, 94
filter, 251
    e-mail, 34
finger, 10
firewall, 124, 251, 302
First Amendment, 110
flag, 274
flame, 302
    war, 302
flames, 112
flaming, 112
FNC, 88
FOAF, 112
folder, 274
FONT, 68
    COLOR, 68
    FACE, 69
    SIZE, 68
food-online.com, 117
footer, 157, 201, 214, 302
foreground, 260
form, 118, 159, 179
FORM, 245
forwarding e-mail, 25
Four11, 166
frame, 40, 303
Framework for Global Electronic
        Commerce, 120
freedom
    of expression, 109
    of speech, 110
freeware, 64, 193, 196, 303
FTP, 38, 47, 195
    get, 195
    put, 195

text-based, 195
transfer mode, 195
full
pathname, 275
text indexing, 184
FWIW, 112

G

Gates, Bill, 99
gatherer, 178, 180
generic top-level domain name, 7
GIF, 36, 107, 222
animated, 299
interlaced, 304
transparent, 310
gigabyte, 303
GII, 120
Global Information Infrastructure, 120
global parameter, 57
glossary, 299
GMT, 15
Good Times hoax, 200
gopher, 3, 38, 47, 93, 303
graphical user interface, 22, 35, 303
graphics file format
TIFF, 36
Graphics Interchange Format, 36, 107
Greenwich Mean Time, 15
grocery shopping, 117
groupware, 122, 303
growth
Internet, 89, 93, 95, 105
Web, 94, 106
guest book, 303
GUI, 22, 35, 303
guide, 141, 162, 311
gunzip, 196
gzip, 196

H

H5, ALIGN, 215
hard link, 280
hash, 303, 305
hashing algorithm, 303
HEAD, 56, 211
header, 157, 201, 211, 303
heading
ALIGN, 77
tag, 75
help file, 197
helper application, 38, 129, 130, 133, 134, 193, 303
versus plug-in, 133
hexadecimal, 63

Hi, 73
hidden
directory, 275
file, 275
hit, 159, 169, 303
hoax, Good Times, 200
home directory, 275
homepage, 41, 52, 303
host, 91
hot button
definition, 38, 303
Destinations, 42
Internet, 42
Lookup, 42
Net Search, 42
Netcaster, 42
New&Cool, 42
People, 42
Software, 42
What's Cool, 42
What's New, 42
Yellow Pages, 42
HotBot, 165
hotlist, 130
HR, 213
ALIGN, 213
NOSHADE, 213
SIZE, 213
WIDTH, 213
HTML, 35
basics, 48
coding style, 50
color, 62
converter, 245
definition, 47
document
background, 64
font, 68
editor, 245
list, 201
style
semantic based, 201
syntactic based, 201
table, 201, 229
tag, *see* tag
attribute, 48
definition, 47
tool
converter, 303
editor, 303
syntax checker, 303
http, 38, 47
HTTP, 97, 106, 128, 304
hyperlink, 38, 43, 44, 303
mailto, 38, 79, 143

hypermedia, 44, 304
hypertext, 44, 304
HyperText Transfer Protocol, 46, 97, 304

I

I, 207
IBM, 91, 100
Idle, Eric, 4
IETF, 106, 304
image
map, 141, 155, 304
transparent, 310
types, 83
IMAP, 29
IMG, 37, 73, 82, 223
ALT, 223
HEIGHT, 37, 84, 223
WIDTH, 37, 84, 223
IMHO, 112
in-line image, 38, 304
inbox, 27, 29
index, 159, 184
index.html, 46
indexer, 178, 184
indexing, 184
full text, 184
Information Superhighway, 88, 89, 106, 107, 304
Infoseek, 113, 163, 165–167
infrared light, 99
Integrated Services Digital Network, 99
Interactive Mail Access Protocol, 29
interlaced
GIF, 304
International Business Machines, 91
Internet, 304
1960's, 91
1970's, 91
1980's, 92
1990's, 93
acronyms, *see* acronyms
address, 304
basics, 88
courses, 94
culture, 108
definition, 88
demographics, 89
firewall, 251
growth, 89, 93, 95, 105
history, 90
Information Superhighway, 88
interesting facts, 89
privacy, 103
Protocol, 88

terminology, *see* glossary
time-line, 90
Internet Engineering Task Force, 106, 304
Internet Explorer, 38, 135, 136, 304
Internet II, 107
Internet Protocol, 304
version 6, 103, 304
Internet Relay Chat, 93, 111
Internet Service Provider, 2, 99, 251
Internet Travel Agency, Inc., 116
Internet Worm, 93, 199
intranet, 107, 124, 304
IP, 88, 304
address, 100, 102, 304
dynamic, 103, 302
static, 103, 253, 309
packet, 97
version 6, 103
IPv6, 103, 304
IRC, 93, 111
ISDN, 99
ISP, 2, 99, 251

J

Jagger, Mick, 12
Java, 89, 94, 130, 304
Oak, 94
Virtual Machine, 304
Java-enabled browser, 304
JavaScript, 130, 145, 304
JavaScript-enabled, 304
Jewel, 12
Joint Photographic Experts Group, 36
JPEG, 36

K

K, 30
KBD, 205
Kbps, 99, 252
keyboard
tag, 205
keys, arrow, 259
keyword search, 170
kill file, 305
Killface, 164, 310
kilobit, 99, 305
kilobyte, 305

L

L.L. Bean, 116
LAN, 91, 99, 305
Lennon, John, 173
LI, VALUE, 219

light
infrared, 99
visible, 99
limitation of liability, 169
line breaking algorithm, 51
line-break, 204
LINK, 67
list, 201, 217
definition, 217, 221
of favorites, 130
ordered, 217
nested, 223
owner, 305
server, 99
unordered, 217, 219
LISTING, 210
LISTPROC, 305
LISTSERV, 305
Local Area Network, 91, 305
location area, 38, 305
search query, 164
login name, 5
Looksmart, 163
lossless compression, 305
lossy compression, 305
Lotus Notes, 121
lurk, 305
lurker, 305
Lycos, 163, 165
Lynx, 38, 305

M

M, 30
Mac, 48
MacCaw, Craig, 99
Macro Virus, 35
Macromedia, 134
Magellan, 163, 165
mail, *see also* e-mail
application, 2, 26
client, 2, 26
program, 26
server, 27, 99
spool, 28
strategy
skim and delete, 33
triage, 33
mail program, Pine, 263
mail-tool, 2
mailbox, 7, 27, 305
mailer, 2, 305
mailing list, 92, 111, 305
public, 12
mailto, 38, 47, 79, 143
mailx, 2

majordomo, 305
MAN, 305
many-to-one, 129
Mastercard International, 114
match, 164, 169
Mbps, 252
McGraw Hill, xvii
MCI, 100
Meeting House, 167
megabit, 100
megabyte, 30, 253, 254, 305
menu bar, 38, 305
Bookmarks, 40
Communicator, 40
Directory, 40
Edit, 40
File, 39
Go, 40
Help, 40
Options, 40
View, 40
Window, 40
Merchant, Natalie, 12
message
board, 251
body, 16
composition, 16
digest, 303, 305
META, 62
CONTENT, 62
HTTP-EQUIV, 62
NAME, 62
Metacrawler, 165, 166
MetaFind, 166
Metasearch, 165
metasearch engine, 162, 163, 165, 305
sample
Metacrawler, 165, 166
MetaFind, 166
Metasearch, 165
SavvySearch, 166
Metropolitan Area Network, 305
Microsoft, 96, 124, 129, 134
Exchange, 2, 121
Internet Explorer, 38
versus Netscape, 134
video, 36
Word, 35, 245
Macro Virus, 35
microwave, 99
MIME, 1, 16, 35, 134, 306
application
postscript, 36
tex, 36
troff, 36
audio

aiff, 36
au, 36
midi, 36
realaudio, 36
wav, 36
file extension
  aif, 36
  aifc, 36
  aiff, 36
  au, 36
  avi, 36
  eps, 36
  gif, 36
  htm, 36
  html, 36
  jpe, 36
  jpeg, 36
  jpg, 36
  mid, 36
  midi, 36
  mov, 36
  movie, 36
  mpeg, 36
  mpg, 36
  png, 36
  ps, 36
  qt, 36
  ra, 36
  ram, 36
  roff, 36
  sgml, 36
  snd, 36
  t, 36
  tex, 36
  tiff, 36
  tr, 36
  txt, 36
  wav, 36
  wrl, 36
image
  gif, 36
  jpeg, 36
  png, 36
  tiff, 36
model, vrml, 36
text
  html, 36
  plain, 36
  sgml, 36
video
  avi, 36
  mpeg, 36
  quicktime, 36
  sgi-movie, 36
mirror site, 45, 305

modem, 99, 146
  speed, 252
  user-to-modem ratio, 252
moderator, news, 306
monospace, 208
Monty Python, 4
Morissette, Alanis, 12
Morris, Robert, 93, 199
Mosaic, 38, 94, 306
Motion Picture Experts Group, 36
mousing over, 43
msvideo, 36
MUD, 306
multi-protocol, 46, 128
Multi-User Dungeon, 306
multimedia, 44, 137, 306
Multipurpose Internet Mail
        Extensions, 1, 35, 306,
    see also MIME
my URLs, 289

N

National Center for Supercomputing
        Applications, 94
National Science Foundation, 92, 93
    Network, 92
navigation, 153
navigational tools, 306
NCSA, 94
nest, 223
nested list, 223
Net Shepherd, 111
Netcaster, 42
netiquette, 20, 112, 306
Netscape, 116, 124, 129, 134, 151,
        163, 306
  Communications, 94, 96, 113
  Communicator, 135
  icon, 38
  versus Microsoft, 134
  Navigator, 38, 39, 94
network
  communication, 96
  convenience, 96
  features, 96
  interface cards, 99
  reliability, 97
  saving, 96
  scalability, 97
  sharing, 96
Network Information Center, 102
Network News Transfer Protocol, 306
Network Service Providers, 100
newbie, 306
news, 38, 47

administrator, 306
article, 299
expired, 302
host, 251
moderator, 306
server, 306
newsfeed, 306
newsgroup, 111, 306
  cross-post, 301
  digest, 301
  follow-up, 302
  moderated, 306
  subscribe, 309
  thread, 310
  unmoderated, 311
  unsubscribe, 311
newsreader, 306
NFSNET, 92
NIC, 102
NNTP, 306
nonbreaking space, 233
Novell Groupwise, 121
NSF, 92, 93
nslookup, 103
NSP, 100
nuclear attack, 91
number, prime, 307

O

O'Riordan, Delores, 48
Oak, 94, *see also* Java
object-oriented programming, 306
obscenity, 110
OC12, 105
Oikarinen, Jarkko, 93
OL
  LI, 217
  START, 218
  TYPE, 218
1,000 bytes, 30
Ono, Yoko, 173
ONSALE, Inc., 117
OOP, 306
open
  architecture, 91
  list, 306
  system, 92
operating system
  UNIX, 252, 273
  VMS, 252
ordered list, 217
organization
  circular, 153
  exploratory, 154
  hierarchical, 155
overkill, 158

P

P, 73
    ALIGN, 217
Pacific Crest Trail, 113
packet, 31, 97, 306
    switching, 91, 98, 306
page, 44, 306, 307, 311
    footer, 201, 214
    header, 201, 211
pages, 44
palette, 301, 307
password, 5, 307
    selection, 6
pathname, 275
pattern matching query, 170, 307
PC, 28
permission, 49
persistent connections, 106
personal page, 52
PGP, 307
physical style type, 207
Pico, 19, 257, 266
PICS, 111
Pine, 2, 257, 263
    composer, 19, 257
    mailer, 19
    Pico, 266
pixel, 79
PKUNZIP, 196
PKZIP, 196
plaintext, 30, 35, 307
Platform for Internet Content
        Selection, 111
plug-in, 38, 129, 133, 193, 307
    versus helper application, 133
    sample
        Acrobat, 134
        Shockwave, 134
PNG, 36, 106
Point-to-Point Protocol, 103, 252, 307
poll, 29
POP, 28, 29
pornography, 110
port, 104
Portable Network Graphics, 36, 106
post, 307
Post Office Protocol, 28
postscript, 36
PPP, 103, 252, 307
pretty print, 50
prime number, 307
Prince William Sound, 176
private
    distribution list, 11
    key cryptography, 307

Prodigy, 251
programming language
    Java, 304
    JavaScript, 304
Project Gutenberg, 307
protocol, 46, 97, 307
prototyping, 157, 307
public
    domain software, 307
    key cryptography, 302, 307
        RSA, 308
    mailing list, 12
public-html, 53
Python, Monty, 4

Q

query, 169, 307
    Boolean, 172, 300
    generalize, 175
    pattern matching, 170, 307
    refine, 176
    semantics, 169, 307
    specialize, 176
    syntax, 169, 307
    too few hits, 175
    too many hits, 176
QuickTime, 36

R

radio wave, 99
README, 197, 215
real-time talk, 123
recursive algorithm, 307
refine, 176
register, 127
    Web page, 308
relative
    pathname, 275
    URL, 57
relevancy score, 169, 180, 308
remote
    host, 254
    login, 89, 187, 308
        command, 192
        introduction, 191
        versus Telnet, 192
render, 37, 308
Request for Comment, 36, 106, 308
resolver, 103, 308
result set, 179, 308
revolving advertisements, 113, 308
RFC, 36, 106, 308
RGB color model, 63, 308

Rivest, Ronald, 308
rm, 277
robot, 180
Romeo, Juliet, 203
root directory, 275
ROT13, 308
rotation 13, 308
ROTFL, 112
router, 99, 107, 308
Roxette, 132
RSA, 308
RTFM, 112

S

s-mail, 2
SavvySearch, 166
scanner, 308
scheme, 46
    file, 38
    ftp, 38
    gopher, 38
    http, 38
    mailto, 38
    news, 38
    telnet, 38
script, 300, 308
scroll bar, 17, 38, 308
search
    backtrack, 182
    Boolean, 172
    breadth-first, 181
    depth-first, 181, 182
    directory
        AOL NetFind, 163
        CNET Search.com, 163
        Excite, 163
        Infoseek, 163
        Looksmart, 163
        Lycos, 163
        Magellan, 163
        Yahoo!, 163, 166
    evaluation, 179
    example, 176
    fundamentals, 167
    generalize, 175
    indexes, 172
    pattern matching, 170
    query, 162
        location area, 164
    refine, 176
    relevancy, 179
    semantics, 169
    specialize, 176
    strategy, 176
    syntax, 169
    tool, 169, 308

directories, 162
  metasearch engine, 162
  search engine, 162
  white pages, 162
white pages
  Bigfoot, 166
  Four11, 166
  WhoWhere, 166
search engine, 10, 115, 159, 161–163, 308
  all-in-one, 165, 299
  component
    evaluator, 178, 179
    gatherer, 178, 180
    indexer, 178, 184
    searcher, 178, 179
    user interface, 178, 179
  meta, 165
    Metacrawler, 165, 166
    MetaFind, 166
    Metasearch, 165
    SavvySearch, 166
  sample
    AltaVista, 165, 166
    AOL NetFind, 165
    Excite, 165
    HotBot, 165
    Infoseek, 165
    Lycos, 165
    Magellan, 165
    WebCrawler, 165
searcher, 178, 179
searching Web, 161
Secure Electronic Transactions, 119
secure server, 117, 308
SELECT, 141
semantic based style type, 201, 202, 308
semantics, 308
sendmail, 26
sequence number, 31, 98
Serial Line Internet Protocol, 252, 309
server, 99, 100, 309
  domain name, 103
  list, 99
  mail, 99
  news, 99, 306
  push, 309
  secure, 117
  Web, 45
server-side include, 309
serving, 52
SET, 119
setting a cookie, 118

7 by 24, 27, 253, 309
SGI-movie, 36
Shamir, Adi, 308
shareware, 64, 309
shell, 273
Shockwave, 134, 137
shopping, grocery, 117
shorthand
  AFAIK, 112
  FOAF, 112
  FWIW, 112
  IMHO, 112
  ROTFL, 112
  RTFM, 112
  YMMV, 112
signature
  digital, 301
  file, 21, 309
Simple Mail Transfer Protocol, 28, 97
Six Flags Theme Parks, 114
sketch, thumbnail, 310
skim and delete, 33
SLIP, 252, 309
SLIP/PPP, 252, 254
SMALL, 206
smiley, 20, 309
SMTP, 28, 97
snail mail, 2, 309
sniffing, 29, 309
social security number, 46
sound, *see* audio
spam, 4, 62, 309
spamdexer, 62
spamming, 4
spell check, 158
spider, 180
splash screen, 67, 145
Sprint, 100
SSI, 309
static IP address, 103, 253, 309
status bar, 38, 309
stemming, 171, 309
Stewart, Martha, 223
storyboard, 309
streaming, 309
STRIKE, 209
STRONG, 203
SUB, 209
Sun Microsystems, 94
SUP, 209
surfer, 44, 309
Surfwatch, 111
switching
  circuit, 91

packet, 91
syntactic based style type, 201, 207, 309
syntax, 309
  checker, 303

T

T1, 105, 252
T3, 105, 252
table, 201
TABLE, 229, 230
  ALIGN, 232
  BGCOLOR, 243
  BORDER, 233
  BORDERCOLOR, 246
  BORDERCOLORDARK, 246
  BORDERCOLORLIGHT, 246
  CAPTION, 241
  CELLPADDING, 233
table
  cells, 231
  divider, 229
TABLE
  CELLSPACING, 233
  HSPACE, 233
  TH, 237
  VSPACE, 233
  WIDTH, 241
tag, 47, 310
  A, 73
  ADDRESS, 204
  B, 207
  BASE, 57
  BASEFONT, 56
  BIG, 207
  BLINK, 210
  BODY, 64
  CAPTION, 241
  CITE, 203
  CODE, 206
  comment, 70
  DD, 221
  deprecated, 210, 301
  DIV, 217
  DL, 221
  DT, 221
  EM, 203
  ending, 49
  FONT, 68
  HEAD, 56
  heading, 75
  Hi, 73
  HR, 213
  I, 207
  IMG, 37, 73, 82, 223

KBD, 205
LI, 217
LISTING, 210
META, 62
OL, 217
P, 73, 217
SMALL, 206
STRIKE, 209
STRONG, 203
SUB, 209
SUP, 209
TABLE, 230
TD, 231
TH, 237
TITLE, 50
TR, 230
TT, 208
U, 209
VAR, 205
XMP, 210
tailgating, 19
target, 310
TCP, 88, 310
TCP/IP, 88, 92, 97, 106, 255, 310
TD, 231
   ALIGN, 239
   BGCOLOR, 243
   BORDERCOLOR, 246
   BORDERCOLORDARK, 246
   BORDERCOLORLIGHT, 246
   COLSPAN, 238
   NOWRAP, 246
   ROWSPAN, 238
   VALIGN, 243
   WIDTH, 242
Teledesic, 100
telephone, 107
Telnet, 38, 47, 187, 310
   escape sequence, 190
   how to invoke, 188
   introduction, 187
   protocol, 188
   versus remote login, 192
terminology, *see also* glossary
   browser window, 38
   search, 169
   Web, 44
TEXT, 65
text
   editor, 16, 19, 257
      basic concepts, 257
      emacs, 257
      Pico, 257
      vi, 257

file format
   HTML, 36
   Plain, 36
   SGML, 36
   transfer mode, 195
TH
   ALIGN, 239
   BGCOLOR, 243
   BORDERCOLOR, 246
   BORDERCOLORDARK, 246
   BORDERCOLORLIGHT, 246
   COLSPAN, 238
   NOWRAP, 246
   ROWSPAN, 238
   VALIGN, 243
   WIDTH, 242
threshold time, 107
thumbnail sketch, 84, 310
tic-tac-toe, 245
TIFF, 36
tilde (~), 275
tiling algorithm, 66, 310
TITLE, 50
title bar, 38, 310
Tomlinson, Ray, 7, 91
toolbar, 38, 310
tooltip, 85, 310
top-level domain name, 7, 8
   generic, 7, 8
Tour de France, 109
Toyota USA, 114
TR, 230
   ALIGN, 239
   BGCOLOR, 243
   BORDERCOLOR, 246
   BORDERCOLORDARK, 246
   BORDERCOLORLIGHT, 246
   VALIGN, 243
traceroute, 31
traffic, 160
transfer mode, 195
   binary, 195, 299
   text, 195
Transmission Control Protocol, 88, 310
Transmission Control
      Protocol/Internet Protocol, 92, 97, 310
transparent
   GIF, 310
   image, 310
tree, 9
   depth, 155
   fan-out, 155

triage, 33, 310
trigger, 199
Trojan horse, 198, 199, 310
TT, 208
tutorial, 141
Tyler, Steven, 12
typewriter, 208

U

U, 209
U. S. Government, 120
UL TYPE, 220
uncompress, 196
under construction, 310
Uniform Resource Characteristic, 185
Uniform Resource Locator, 38, 44, 128, 310
United States, 8
universal service, 3, 310
University of Washington, 263
UNIX, 252, 273, 310
   basics, 273
   command, 273
      cd, 276, 281
      chmod, 280, 281
      cp, 278, 281
      date, 275
      fg, 260
      ls, 274, 281
      man, 273, 281
      mkdir, 276, 281
      more, 278, 281
      mv, 277, 281
      pwd, 281
      rm, 277, 281
      rmdir, 277, 281
   directory, 274
   file, 274
   flag, 274
   foreground, 260
   introduction, 273
   mkdir, 53
   path, 275
   pathname, 275
   permission, 278
   prompt, 273
   shell, 273
UNIX to UNIX Copy, 92
unordered list, 217, 219
URC, 185
URL, 38, 44, 106
   absolute, 57
   relative, 57
USENET, 92, *see also* newsgroup
User Network, 92

userid, 5, 299, 311
UUCP, 92
UW, 257, 263

V

vacation program, 33, 311
vanity plate, 6
VAR, 205
VC, 122
Vedder, Eddie, 12
Veronica, 198
version control, 122
vi, 257
video, 123
    teleconferencing, 122
videoconferencing, 96, 122, 311
    audio, 123
    black and white video, 123
    chat, 123
    color video, 123
    whiteboard, 123
Virtual Monitoring System, 252
virtual reality, 311
virus, 36, 187, 198, 311
    detection, 193
visible light, 99
VLINK, 67
VMS, 252
VRML, 36

W

W³, 43
W3, 43
W3C, 106, 312
WAIS, 311
walknet, 193

Wall Street Journal Interactive
    Edition, 116
WAN, 305, 311
Warner Bros., 130
way-station, 311
Web, 43, 311
    basics, 127
    bookmark, 40, 130
    browser, 2
        definition, 37
    cache, 129
    client, 44, 100
    cookies, 129
    directory, 162, 311
    favorites, 130
    find, 42
    growth, 94, 106
    guide, 162, 311
    manager, 45
    page, 44, *see also* tag
    pages, 44, 307, 311
    personal preferences, 129
    presentation, xvi, 44, 145, 311
    reload, 41
    server, 10, 45, 46
        definition, 52
    site, 45, 311
    surfing, 43
    writing styles, 136
Web server, 311
WebCrawler, 165
Webmaster, 44, 311
white pages, 10, 162, 163, 166, 311
white space, 49
whiteboard, 123
WhoWhere, 166
Wide Area Information Service, 311
Wide Area Network, 311

Windows PC, 48
wink, 20
WinQVT, 188
woodworking, 62
Word, 245
working directory, 276
world, readable, 53
World Wide Wait Problem, 87,
    105–108
World Wide Web, 43, 93, 127, 311
    Consortium, 106, 312
    Problem, 312
worm, 180, 198, 199, 312
Worm, Internet, 93, 199
writing
    genre, 136
        biography, 137
        business exposition, 139
        guide, 141
        tutorial, 141
    style, 136
WWW, 43, 127, 311

X

XML, 302
XMP, 210

Y

Yahoo!, 10, 94, 113, 163, 166, 312
Yang, Jerry, 94
yellow pages, 166
YMMV, 112

Z

Zimmermann, Phil, 307
zip, 196